Table of Contents

List of Figures

List of Tables

NATION BUILDING MODELING
AND RESOURCE ALLOCATION
VIA DYNAMIC PROGRAMMING

I. Introduction

Historically, nation-building attempts by outside powers are notable mainly for their bitter disappointments, not their triumphs[73]

Pei and Kasper

1.1 Motivation

In the history of the United States (US) there have been over 200 instances where the armed forces have been used abroad; since 1900, 18 of those instances have been nation building operations [73]. Most recently the operations in Afghanistan and Iraq are clear examples of the full spectrum of operations, where major combat operations became nation building operations. Considering these 19 instances, Iraq and Afghanistan not included, only four (Japan, West Germany, Grenada, and Panama) were still a democracy 10 years after the operations concluded. The recent track record for this metric of long term success is less than desirable.

The two recent instances of these type of operations have taken place in Afghanistan (2001-ongoing) and Iraq (2003-2011). In terms of the military support that has been provided, it has come in the form of major combat operations and nation building operations. The frequency, duration, and complexity of these operations describe a perceived threat to not only US national security but security in all regions of the world, most recently demonstrated by France and its effort in Mali (2013-ongoing).

1

The modern concept of nation building can be defined as "the use of an armed force as part of a broader effort to promote political and economic reforms with the objective of transforming a society emerging from conflict into one at peace with itself and its neighbors" [38]. When evaluating major combat operations (MCO), the measurements are more clearly defined and easier to measure. The kinetic nature of the operations have distinguishable outcomes (i.e. number of forces killed, number of objectives secured) which are typical incorporated in today's combat models. From analysis of nation–building operations it is clear to see that the metrics to evaluate these operations are not well defined. However, the frequency of operations is increasing and the costs are becoming more severe.

Most current research is focused on the first three phases of Joint operations (Deter, Seize the Initiative, and Dominate) even though history has shown that the Phase IV (Stabilize) operations can be the longest and most critical when setting the conditions for long term peace and stability. Therefore, it is crucial to allocate resources for these type of operations which produce the best result with the least cost; an optimal solution.

1.2 Research Contribution

This dissertation demonstrates how a political and social science concept such as nation building can be modeled as a system of differential equations and analyzed using dynamic programming. This is a multi-step approach; developing a set of indices to define the state, using a system of differential equations to define the transfer function, and finally developing improved policies using dynamic programming. Thus, the objective is to determine how US resources can be allocated in a nation building operation to set the conditions for long term success while minimizing the costs associated with expending those resources. This is accomplished through five contributions:

1. Develop a methodology which creates indices to capture the "state" of a nation. This method is novel and innovative in that it makes use of open source data and is adaptable to the set of available data. This will be accomplished using the DIME–PMESII (defined in Chapter II) paradigm.

2. Development of a model that accounts for the evolution of the state of a country and what influence PMESII states have on one another; as well as the impact of external influences, namely DIME inputs.

3. Development of a near–optimal policy using approximate dynamic programming, based upon the resources which the US can provide in nation building operations.

4. Development of a model that accounts for the evolution of the state of a country and what influence PMESII states have on one another; as well as accounting for the impact of external influences, namely DIME inputs and enemy influences as measured by stochastic shocks or events.

5. Development of a near–optimal policy using approximate dynamic programming by implementing shocks based on the number and severity of attacks and laying the framework for consideration of the probabilistic behavior of enemy shocks or events.

This is demonstrated through examples using data from operations in Iraq.

1.3 Organization of Dissertation

The remainder of this dissertation is organized into six chapters. Chapter II contains a detailed literature review, Chapter III describes the system of differential equations model, Chapter IV describes the dynamic programming solution, Chapter

V describes the dynamic programming solution with enemy events, and Chapter VI describes future research efforts.

II. Literature Review

This chapter examines the literature which applies to the creation of the model and solution methodology. Each area focuses on a concept and how that concept was applied in this methodology. The literature review provides the reader with a succinct but thorough overview of areas of study that are directly related to addressing the research question. In following chapters, literature that directly relates to a specific chapter is presented as well.

2.1 DIME–PMESII Paradigm

The instruments of national power (Diplomatic, Informational, Military, and Economic), sometimes referred to as the elements of national power, or categorized under the acronym DIME , are tools at the disposal of the US to aid in achieving its national strategic objectives. All four instruments may be applied or a subset of the four; they are not necessarily mutually exclusive events. Operational planning is described in terms of six interrelated operational variables (*Political, Military, Economic, Social, Information, and Infrastructure* or PMESII). The PMESII variables describe the military aspects of the operational environment as well as the influence of the population on the operational environment. As a result, they provide a view which emphasizes the human aspects of the operational environment. A thorough understanding of PMESII helps to appreciate how the military instrument complements the other instruments of national power [36:1-5]. The DIME-PMESII paradigm considers the the relationship between the two concepts. A description of each instrument and operational variable follows.

2.1.1 Diplomacy (DIME).

The diplomatic instrument is the primary means employed by the Department of State (DOS) to engage other states to advance the values, interests, and objectives of the US [34:I-9].

Diplomacy takes on many different actions, including, but not limited to: providing mentors to a fledgling government, providing an interim structure to aid during a transition period, providing and conducting elections, and providing support to legitimize a government in the eyes of other states.

Assessing and determining the level of diplomatic assistance is a complicated task that may be controversial. The amount of support is based on elections and diplomatic support levels, a government run by an external government is a value of 1, shared support with the host nation is a value of 0.5, and completely autonomous elections run by the host nation is a value of 0 for the purposes of this research.

2.1.2 Informational (DIME).

The informational instrument deals with both the protection of information and the distribution of information. According to JP 1-0, the uses of the informational instrument are

> ...processes and efforts to understand and engage key audiences to create, strengthen, or preserve conditions favorable to advancing national interests and objectives through the use of coordinated information, themes, messages, and products synchronized with the actions of all instruments of national power. [34:I-9]

The informational instrument is perhaps one of the most complex instruments to manage because it is so difficult to measure. The ability to determine the impact or calculate the reception of a message in a contested area is often unknown. For this reason the informational instrument is not included in this research effort.

2.1.3 Military (DIME).

The military instrument is perhaps the most well–known, often used aspect of DIME. The US military can take on a wide range of operations in order to support the US national strategic objectives. In this instance the focus is on nation building and related operations. In these operations the application of the military instrument continues well after the completion of a conflict, humanitarian operation, or other type of operation. The wide range of operations the military can conduct makes it a valuable commodity in the application of DIME. In this research, the military instrument is defined as the number of troops (per hundred thousand per month) on the ground in support of a US mission.

2.1.4 Economic (DIME).

The typical employment of the economic instrument in nation building is through aid packages and assistance to the nations economy. This is to aid in making a self–supportive nation, when aid can be curtailed. In this research the economic instrument is defined as the amount of economic support in dollars (per billion per month) provided to the nation.

2.1.5 Political (PMESII).

The political variable is a description of the distribution of responsibility and power across all levels of government. There may be conflicting political groups and each may interact with the US or multinational force differently. Understanding the unique circumstances that motivate and drive these groups requires an understanding of all the relevant partnerships and their interactions– *political, economic, military, religious, and cultural* [36:1-5,6].

2.1.6 Military (PMESII).

The military variable describes the military capability of all armed forces in a specific operational environment. The armed forces of a state may include the role of providing both internal and external security. Additionally, influencing the military variable are paramilitary and guerilla forces. The organization's ability to field capabilities and use them locally, regionally, or globally is one way of assessing the military variable. These capabilities include: 1) Equipment, 2) Manpower, 3) Doctrine, 4) Training levels, and 5) Resource constraints [36:1-6].

2.1.7 Economic (PMESII).

The economic variable deals with the behaviors of individuals and groups pertaining to the production, distribution, and consumption of resources. The influence of industry, trade, development, finance, policies, capabilities, and legal constraints will play a significant part in the behaviors associated with economics. Factors associated with changes in the economic environment may include investments, price fluctuation, debt, and the existence of black markets [36:1-6,7].

2.1.8 Social (PMESII).

The society within a operational environment is the social variable. A society is "a population whose members are subject to the same political authority, occupy a common territory, have a common culture, and share a sense of identity" [36:1-7]. As with many factors the attributes of a society may change over time, leading to a split within a society. The societies actions, opinions, or influences should be considered within the social environment, as they can have an effect on the mission [36:1-7].

2.1.9 Information (PMESII).

The information variable describes the information environment, which is the network of people, organizations, and systems that collect, process, disseminate, or act on information. Not all states have a complex telecommunications network to share information, but nonetheless the information will be shared through less sophisticated methods [36:1-8]. Due to the complex nature in measuring the effectiveness of information, it was not be considered in this research.

2.1.10 Infrastructure (PMESII).

The infrastructure variable describes the basic facilities, services, and installations required for a society to function. This also includes technological advances and development which can be applied to both civil and military purposes [36:1-8].

2.1.11 DIME-PMESII Summary.

While each operational environment is different and constantly evolving over time, the PMESII variables are used to help understand this complex adaptive environment, while the DIME inputs are used to understand the actions conducted. This DIME-PMESII term is typically used to describe operations [53]; specifically nation–building operations. For this reason the system of differential equations is modeled using this DIME-PMESII paradigm. The exclusion of the informational instrument and operational variable leaves a slightly abridged version of the paradigm, which is described as DME–PMESI or PMESI-DME from here forward.

2.2 Inverse Problems

The goal in solving several types of problems is to determine the set of parameters which describe the system and the laws and principles relating the values of the parameters to the results of measurements. When some information is known about the values of the measurements, a theoretical relationship can be used to infer information on the values of the parameters. When the problem is posed in this manner, it is called an inverse problem. In inverse problems, the data are results of the measurements and the unknowns are the values of the parameters [92]. Partial information is given or known about a state function, $x(t)$ and the goal is to infer something about the laws governing state evolution, values of constant parameters, values of exogenous functions which characterize the system, or values of boundary conditions at certain points in time [45].

Tarantola and Valette propose that all problems, inverse or not, to be stated as well-posed problems are formulated as follows:

1. We have a certain state of information available on the values of the data set [92].

2. We also have a certain state of information on the values of the unknowns [92].

3. We have a certain state of information concerning the theoretical relationship that exists between the data and the unknowns [92].

4. Which is the final state of information on the values of the unknowns resulting from the combination of the three preceding states [92]?

Many experiments contain a finite amount of data in which one can reconstruct a model with infinitely many degrees of freedom. The result is an inverse problem is not unique in that there are many models that can explain the data. According to Tarantola and Valette, inversion really consists of two steps, from the traditional inverse problem there is the true model (m) and data (d). From the data, d an estimated model $(\hat{m})$ is constructed, this is an *estimation problem*. Additionally, the relationship between the estimated model, $\hat{m}$ and the true model, m must be investigated. This is called the *appraisal problem* [89]. The notion of this division of problems is illustrated in Figure 1. The estimation problem is typically solved by fitting the model to the data, by letting the ith data element d_i be related to the

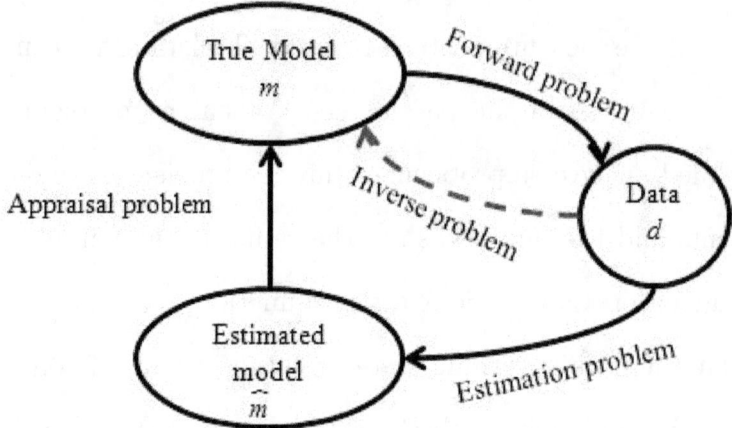

Figure 1. A problem divided into a forward problem, estimation problem, and appraisal problem for finite data sets, adapted from [89:389]

model m through the following relation

$$d_i = G_i(m) \tag{2.1}$$

where $G_i(m)$ is a nonlinear function and $G_i(m) \mapsto d_i$. The data fitting can be accomplished by minimizing the difference between the real data, d_i and estimated data $G_i(\hat{m})$ through a least-squares fit

$$S(\hat{m}) = \sum_i (d_i - G_i(\hat{m}))^2 \tag{2.2}$$

as a function of the estimated model $\hat{m}$[89].

The use of inverse problems in this research is to infer information about the dynamics of the system through a priori knowledge of the exogenous system variables, much like Lanchester equations in Section 2.3.

2.3 Lanchester Equations

Within military applications, one of the most famous treatments of inverse problems are the Lanchester equation solutions, published in F. W. Lanchester's *Aircraft in Warfare: The Dawn of the Fourth Arm in 1916*. While much of the focus of Lanchester's book centers on aircraft and their emerging use in World War I, his major contribution was to offer a set of differential equations to model combat power for both enemy (x) and ally (y) strength, by using existing data. Beginning with aerial combat he developed the Linear Law (unaimed fire)

$$\frac{dx}{dt} = -Axy$$
$$\frac{dy}{dt} = -Byx \tag{2.3}$$

where attrition is proportional to the attrition coefficients (A, B) and the size of both forces (x, y). This is associated with area fire such as indirect fire. To deal with direct or aimed fire, he developed the Square Law (aimed fire)

$$\frac{dx}{dt} = -Ay$$
$$\frac{dy}{dt} = -Bx \tag{2.4}$$

where attrition is proportional to the strength (x, y) and effectiveness (A, B) [55]. Later Lanchester broadened his equations to apply in other types of conflicts. As an inverse problem, the coefficients are determined through knowledge of the data at time t. Lanchester's work continues to serve as the basis and motivation for much research, to include this research.

Over time, several researchers have used Lanchester equations on prominent battles as the data became available. In addition to using new data, they also applied the equations to different types of conflicts, evaluated several of the parameters, and utilized multiple methods to solve the parameters. Bracken generalized Lanchester equations to model the Ardennes campaign in World War II, where he considered the performance of either opposing force at a point in time, with tactical parameters and attrition rates. Bracken solved for these parameters by implementing a *brute–force* method through a constrained grid search [22]. Extending Bracken's work, Fricker examined the same Ardennes campaign, but used a linear regression technique [41]. Clemens analyzed the same data set utilizing a nonlinear fit with the Newton-Rhapson algorithm [30]. Helmbold makes use of the Newton-Rhapson algorithm while examining the square law with scheduled reinforcements, as a direct problem and as an inverse problem [45]. Lucas and Turkes applied a response surface methodology to the Ardennes data set and solved for the parameters by regression through the origin. This method allowed them to use a contour plot and visually assess the optimal point for the parameters. Lucas and Turkes also advanced the idea of using R^2 when using linear regression to compare models using weighted data [56]. Previous methods had primarily focused on the sum of squared residuals or sum of squared errors (SSE).

The conjugate gradient method is used by Chen to determine the coefficients for time dependent attrition in the nonlinear Lanchester square law inverse problem

$$\frac{dx_1(t)}{dt} = - D(t; x_1, x_2)x_2(t) + \frac{dR_1(t)}{dt}, \qquad t > 0; x_1(0) = x_{1,0} \qquad (2.5)$$

$$\frac{dx_2(t)}{dt} = - A(t; x_1, x_2)x_1(t) + \frac{dR_2(t)}{dt}, \qquad t > 0; x_2(0) = x_{2,0} \qquad (2.6)$$

where

$$A, D \text{ are force dependent attrition coefficients}$$

$$R_1, R_2 \text{ are the total reinforcement}$$

$$x_1, x_2 \text{ are the estimated force strengths}$$

by making use of the observed force strength data. In order to numerically solve these equations the fourth–order Runge–Kutta method was used. Chen found this method to be advantageous because there was no prior knowledge required to solve for the unknown parameters. This method allows for an arbitrary initial starting point [29].

The application of Lanchester Equations has been documented in other areas as well, such as irregular warfare. Schaffer, for example, used Lanchester Equations to model guerilla warfare and asymmetric engagements while employing and operationalizing an array of variables and coefficients (representing weapons strength, discipline, morale, *etc.*) to model an insurgent force in Phase II of an insurgency [84]. Richardson used a system of differential equations to model the arms race and instability of nation states based upon the current levels of its neighboring and/or *menacing* states [78].

The use of discrete dynamical systems (DDS) by Fox [40] continues the work of Lanchester equations and proposes a model for insurgency and counter-insurgency warfare using the paradigm, Future = Present + Change. Fox's use of DDS is based upon the mathematical properties they provide. His paradigm is similar to an Euler step where the system transitions to future states based upon its current state and the current rate of change.

In 2010 Lukens [57] describes the DOD desire to expand irregular warfare (IW) modeling to inform program decisions and proposes a model to replicate the dynamics of IW. This model extends basic Lanchester equations to account for four factors (friendly forces, enemy forces(terrorists and insurgents), host nation forces, and the population. Lukens describes modeling the elements of Power (DIME) in IW, noting that the modeling these elements is extremely complex and not recommended in a basic warfare dynamics model. Lukens provides a rough framework to look at DIME but does not include it in his model.

In 2011 Atkinson, Gutfraind, and Kress [5] incorporates the concept of foreign intervention in Lanchester models, specifically in the setting of an armed revolt. This adds the aspect of an external force in to the equation, whereas models such as Lukens [57], Fox [40], and others only consider internal forces. This external force is representative of nation building operations and demonstrated in recent revolts (Libya, Syrian, and Afghanistan).

2.4 Mean–Field Theory

Many problems involve a large number of independent variables where the exact calculation of such a problem is infeasible. In order to solve these problems, efficient approximation techniques are needed in order to better understand their dynamics. The method of using a Mean–Field Equation (MFE) to approximate these dynamics is an efficient approximation method to aid in solving problems dealing with uncertainty and complexity [72:ix,1].

In this method, the values of the variables to be examined are replaced with the MFE. The variables of the dynamical system are used to determine some mean value; this is accomplished through an equation that provides the mean–field simplification. This allows the focus to be placed on one variable at a time by effectively holding the others constant. To consider this intuitively, envision a problem with multiple variables and only one is not represented by its mean value. This leaves the one free

15

variable independent of the others, thus creating the ability to calculate the value of the free variable. The process is then repeated in the same fashion over all of the remaining values. Persson, Claesson, and Nordebo use this technique to conduct discrete adaptive filtering using a mean–field algorithm to minimize the Wienr-Hopf equations in a least–squares sense to produce comparable results without transient behavior and to facilitate abrupt system changes [74].

The mean–field equations can be the mean value or an approximated probability distribution to represent the unknown variables. With a large number of variables that exhibit nonlinear behaviors fitting them with a nonlinear least–squares model is an effective method [74]. This concept was shown using an epidemic model based upon a system of differential equations by Kleczkowski and Grenfell [51].

This method injects a portion of generality into the process which still accounts for the noise in the system, yet simplifies the problem by using a constant in the place of a changing variable. This method replaces some of the stochastic elements with deterministic elements, resulting in a stochastic system represented through its deterministic equivalent.

This method is not without error, as the number of estimated variables reduces the overall confidence level of the result by one degree of freedom with each estimated variable. This does not indicate inaccuracy, but rather that the end result will be an overall estimation of the system based upon the previous interactions. It may downplay the effect of outliers in the generalization, but it does account for them. The error in the system is expected to be normally distributed. This is important to the least–squares fitting aspect and the independence of the variables in the system. This method replaces an infinite dimension system with several dependent variables, with a series of independent variables in a finite dimension system, thus reducing the overall complexity of the problem and allowing us to understand the dynamics of complex problems.

2.5 Dynamical Systems

When creating models, a real world system to study is identified, all the aspects of that system are studied, and assumptions are made when and where they are needed. After studying the system, it is often translated into a mathematical relationship which can be modeled. A knowledge of mathematics is used to conduct analysis on this system in order to solve the complicated interrelationships that exist in the real world system. This solution then translates the knowledge gained from the model back to the real world system. Dynamical modeling is the science of modeling real world phenomena as it changes over time [82:3].

According to Boccara and Meiss, the definition of a dynamical system is a set or system of equations whose solution describes the evolution or trajectory of the state, as a function of a parameter (time), along a set of states (phase space) of the system [82:105-106] [19:11]. The theory behind dynamical systems is primarily concerned with the qualitative properties of the system dynamics and gaining an understanding of the asymptotic properties, as $t \to \infty$. A typical dynamical system is comprised of a phase space, $\mathcal{S}$, whose elements represent all possible states for the system; a time parameter, t, which may be discrete or continuous; and an evolution rule (a rule that governs the transition of states from t_i to t_{i+1} based upon knowledge of the states at prior times) [19:105-106]. A dynamical system is characterized according to these three elements. Systems with both discrete time space and time variables are often considered mappings. When the evolution rule is deterministic then for each time, t, it is a mapping from phase space to phase space

$$\varphi_t : \mathcal{S} \to \mathcal{S} \tag{2.7}$$

so that $x(t) = \varphi_t(x_0)$ indicates the state of the system at time t that begins at x_0. The value of t is assumed to only take on values in some allowed range, the set of nonnegative real numbers $\mathbb{R}^+$ and the initial value of $t = 0 \Rightarrow \varphi_0(x_0) = x_0$ [82:106].

Dynamical systems can be modeled by a finite number of coupled first-order ordinary differential equations

$$
\begin{aligned}
\dot{x}_1 &= f_1(t; x_1, \ldots, x_n; u_1, \ldots, u_p) \\
\dot{x}_2 &= f_2(t; x_1, \ldots, x_n; u_1, \ldots, u_p) \\
&\vdots \\
\dot{x}_n &= f_n(t; x_1, \ldots, x_n; u_1, \ldots, u_p),
\end{aligned}
\tag{2.8}
$$

where $\dot{x}_i$ is the derivative of x_i with respect to time, t, and the set of variables $u_1, u_2, \ldots, u_p$ are control variables required for that system. The variables $x_1, x_2, \ldots, x_n$ are the state variables and represent the memory the dynamical system has of its past. In order to write these systems in compact form, vector notation is generally used. First, the vectors are defined as

$$
x = \begin{bmatrix} x_1 \\ x_2 \\ \vdots \\ x_n \end{bmatrix}, \quad
u = \begin{bmatrix} u_1 \\ u_2 \\ \vdots \\ u_p \end{bmatrix}, \quad
f(t, x, u) = \begin{bmatrix} f_1(t, x, u) \\ f_2(t, x, u) \\ \vdots \\ f_n(t, x, u) \end{bmatrix}, \tag{2.9}
$$

and then rewritten as a compact first-order vector differential equation

$$\dot{x} = f(t, x, u). \qquad (2.10)$$

This is the state equation where x is the state and u is the control. Another equation

$$y = h(t, x, u), \qquad (2.11)$$

may define an output vector comprised of variables of particular interest in the analysis of the system. The two together form the state space model or state model. Mathematical models of finite dimensional systems are not always developed in the form of a state model. However, physical systems can thoroughly be modeled in this form by carefully selecting the state variables [50:1-4].

While nonlinear systems are often more accurate models of real world systems than linear models, many of the linear models are actually linearizations of nonlinear models because it is often difficult to find a closed form solution of a nonlinear system. By using the appropriate techniques it is possible to determine qualitative behaviors of the solutions of a nonlinear system which is desired [82:367].

2.6   Reconstruction Operations

One attempt to model the dynamics involved in reconstruction operations was using systems dynamics modeling techniques to simulate the establishment of public order and safety by Richardson. The purpose was to help decision makers by providing insight regarding the possible policy alternatives presented to them. The main idea is to take complex problems and break them down into manageable subproblems, then aggregate assumptions about the simpler questions to estimate answers for the

19

larger complex problem. This was demonstrated in a notional example at a national level [77].

Robbins then advanced the model by instituting a sub-national, regional level approach. This allowed the user to concentrate on potentially troublesome regions, by providing information specific to the dynamics within that AOR. The results help the user understand the significance of the dynamic relationship of forces involved during nation–building and potentially gain insight to the successful completion of the nation–building mission [79]. This model eventually was re-engineered by Air Force Research Laboratory–Rome Laboratory (AFRL-RL) to become the National Operational Environment Model (NOEM) currently maintained by the same organization.

The application of goal programming was conducted by Bang to formulate the Coalition Operation Planning Model which was based upon three different submodels: the Coalition Mission-Unit Allocation Model (Shortest Path), the Coalition Mission-Support Model (Network Flow), and the Coalition Mission-Unit Grouping Model (Quadratic Assignment). This method was applied to notional humanitarian assistance scenario and showed that many of the decisions were directly influenced by the political nature of the coalition and the framework provided by the political situation [7].

Tauer, Nagi, and Sudit [93] formulated a simplified version of the model by Richardson [77] as a markov decision process. Tauer *et al* used a Reduced Approximate Linear Program for the H-neighborhood around an initial state x_0, assuming a given expert's policy π_E ($RALPH_E^H$). This was used to model the transition between population classes (Unemployed, Private, Government).

A goal programming project scheduling approach was conducted by Chaney to prioritize and schedule activities to maximize the impacts in nation–building. This was established through three goals: 1) restore essential services in a timely manner, 2) distribute employment equally throughout the state, and 3) meet standards for sustainable income in each region. This was applied in a notional scenario, and showed how to schedule activities to meet the three goals while still meeting the intent of the initial response. Chaney presented three main points in this work: 1) consider economic impacts of reconstruction activities, 2) quantitative project scheduling techniques can be applied to nation–building, and 3) the establish of these techniques adds defensibility to the plan and can uncover potential shortfalls [27].

2.7 Network Models

A social network analysis study was conducted by Bernardoni using Ronald Burt's structural hole technique to facilitate nation–building in failing and failed states. Bernardoni applied Burt's technique at a national level to identify the structural gaps within a failing state by focusing on techniques that link professional and government community individuals [9].

Arney and Arney [4] use a large scale system of differential equations and networks to model counter-insurgency and coalition operations in stages. The network model describes the collaboration link between nodes while the system of differential equations provides the metrics to evaluate operations. While the aspect of external forces is applied, the metrics are largely based on populations of groups within the model.

2.8 Classification Models

With the number of failing or failed states on the rise, the ability to determine the indicators which lead to a failed state and identify states which are failing is a desirable feature. Nysether used factor analysis to identify the indicators and then apply discriminant analysis using the identified factors to classify states as stable, borderline, or failing. This was applied using opensource data for 200 countries with 167 variables. This research is useful in identifying states which may require future nation–building [68].

Understanding the factors which lead to war termination was researched by Robinson through the use of binary and multinomial logistic regression techniques. Robinson found that duration of conflict was the most relevant factor in predicting the winner of conflict and total casualties was the most relevant factor in predicting the manner in which an interstate war ends. This was examined in analysis of 19th and 20th century data [80].

Using the same methods as Nysether and Robinson as well as Canonical Correlation and Principal Component Analysis (PCA), Tannehill develop a mathematical model to forecast instability indicators in the Horn of Africa region using 54 variables over 32 years of observations. This model used indicators such as battle deaths, refugees, genocide deaths, and undernourishment to forecast instability. Tannehill found that a four–year forecast was possible while maintaining or improving the forecast error rate. This demonstrated the feasibility of longer term predictive models which would allow policy makers more time to develop plans [91].

In 2007, the Center for Army Analysis CAA initiated the Forecast and Analysis of Complex Threats (FACT) study. This study looked at predicting the potential for future conflict in select nation–states. The study found 13 features to measure and scaled the features on a $[0, 1]$ scale using the Euclidian distance between a nation–states forecasted future and all other nation–states pasts, both points in the 13 dimensional feature space. A PCA was conducted in an attempt to reduce the dimensionality of the data. The components then provided a proxy for similarity between states. A forecast was then generated using a Weighted Moving Average

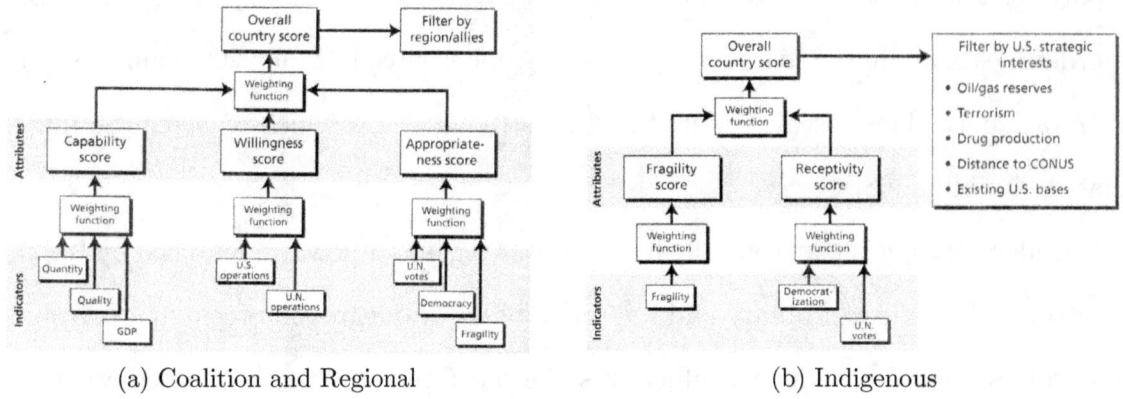

(a) Coalition and Regional   (b) Indigenous

Figure 2. RAND Study Models[59:98,115]

(WMA) and used both the k-Nearest Neighbor (KNN) and Nearest Centroid (NC) algorithms to classify future features. The study found that KNN performed better than NC with 85% or greater accuracy in all test cases. The methodology was adopted for use under the premise that it is predictive rather then prescriptive as described by Shearer [87].

The RAND Arroyo Center conducted a study looking at the strategic elements to build partner capacity for stability operations in nations around the globe. The study focused on the elements which would align the security cooperation efforts of the US and building partner capacity. As part of this study they created two models; the Coalition and Regional model and the Indigenous model, shown in Figure 2.

The Coalition and Regional model was used to assess the capability of nations to be partners in stability operations: 28 countries fell into the high capability category, 5 of which were considered preferred. The study concluded that some high capability countries are either unwilling to participate and/or are inappropriate for such operations. The preferred countries were Argentina, Czech Republic, Hungary, Poland, and South Africa [59:111].

The Indigenous model assessed how fragile a state was and the threat posed if they deteriorated or collapsed. The study found that this model also listed 28 of the

31 countries listed in the Fund for Peace's Failed States Index. Out of the 28, 16 were candidates based upon the US having two or more strategic interests with that country. The 16 are Afghanistan, Columbia, Cuba, Egypt, Indonesia, Iraq, Kuwait, Mexico, Nigeria, Pakistan, Peru, Qatar, Saudi Arabia, Turkey, United Arab Emirates, and Venezuela. Several of these nations are receiving aid already or are considered ineligible because of the current government in place [59:122-123].

The study concluded that it would be beneficial for the US to develop a selective strategy for partnership that nests with the security of the nation and the national military strategy [59:123-124].

In 2009 Abdollahian, Nicholson, Nickens, and Baranick [1] provided the Formal Stabilization and Reconstruction Operations Model (FSROM) using a system of differential equations and seeming unrelated regression estimation (SURE) determine the optimal degree of foreign aid, multilateralism (number of nations participating), and operation length. The model was tested using instances of nation building operations from post World War II through operations in Afghanistan and Iraq with mixed but promising results. The premise of the model is simple gains and losses, where comparisons are made to pre-conflict levels. Several of the equations are related to the size of the guerilla force, which indicates an underlying population model. The authors elude to the ratio of troops to guerrillas as a contributing factor to the success/failure in some of the case studies, yet overall conclude there is no "magic ratio". Two observations are pointed out in the work. First, the resolution of the data, both spatial and time is not adequate,. Secondly, that more detailed (stabilization and reconstruction) factors were necessary to identify the important aspects of policies and developing courses of action.

The FSROM model attempts to capture three aspects of stabilization and reconstruction operations. None of the three are controllable by the US. The US certainly has an impact on the foreign aid, size of the coalition, and the length of the operation but all of these aspects are outside the purview of the US. While providing analysis through these variables may provide insight to the problem, this model is still lacking the prescriptive nature described by Shearer [87]. Any adjustments in policy based on analysis from this model requires a multilateral effort.

In 2014 King [46] looked at classifying, predicting success, and estimating forces required to conduct counter insurgency operations. While this research makes uses of multiple aspects of DIME, the key output is the number of external forces to maximize the probability of success in counter insurgency.

2.9 Insurgency and Counterinsurgency Models

In 2008 Blank *et al*, developed a dynamic model of insurgency using Lanchester equations and Iraq war data. The model proposes a system of differential equations

$$
\begin{aligned}
\frac{dI}{dt} &= (r_i - \gamma_c)C \\
\frac{dC}{dt} &= (r_c - \gamma_i)I,
\end{aligned}
\tag{2.12}
$$

where

I is the number of insurgent attacks on the coalition

C is size of the coalition

r_i is the recruitment rate of the insurgents

r_c is the recruitment rate of the coalition

γ_i is the combat effectiveness coefficient of the insurgents

γ_c is the combat effectiveness coefficient of the coalition.

The general solution to the system of differential equations is then used to plot the phase portraits of the system and deduce information based upon four cases: 1) the coalition increases in size and the number of attacks by the insurgents increases, 2) the size of the coalition decreases and the number of attacks by insurgents decreases, 3) the coalition increases and the number of insurgent attacks decreases, and 4) the coalition decreases and the number of insurgent attacks increases [18].

The relevance of the case is dependent upon the net recruitment rates $(r_i - \gamma_c)$ and $(r_c - \gamma_i)$ of the coalition and insurgents as well as the combat effectiveness of both sides. Using these plots the case where there is no coalition presence and the insurgent attacks are zero, the system is unstable, implying there is no amenable solution that leads to stability [18].

A nation–building model investigating the assimilation of different ethnicities into a single nation was developed by Yamamoto. This model was derived from the system of differential equations in the Deutsch Model for Nation–Formation by Karl Deutsch. Yamamoto derived two models and applied them to the Philippines [102].

The Modernism model is predicated upon the belief that a single underlying population (U) will mobilize into two different groups, assimilated (N) and differentiated (H). The Modernism model is formulated as

$$
\begin{aligned}
\frac{dN}{dt} &= gN + \alpha m U \\
\frac{dH}{dt} &= gH + (1 - \alpha)mU \\
\frac{dU}{dt} &= gU - mU,
\end{aligned} \tag{2.13}
$$

where

g is the natural population increase rate

m is the mobilization rate

α is the rate of integrating into the assimilated group (N)

$\alpha, m \in (0, 1]$.

The Historicism model is predicated upon the belief that underlying population (U) is composed of two groups (Q, R) which will mobilize into the assimilated $(Q \mapsto N)$ and differentiated $(R \mapsto H)$. The Historicism model is formulated as

$$
\begin{aligned}
\frac{dN}{dt} &= gN + \alpha m Q \\
\frac{dH}{dt} &= gH + (1 - \alpha)mR \\
\frac{dQ}{dt} &= gQ - \alpha m Q \\
\frac{dR}{dt} &= gR - (1 - \alpha)mR.
\end{aligned}
\tag{2.14}
$$

where

g is the natural population increase rate

m is the mobilization rate

α is the rate of integrating into the assimilated group (N)

$\alpha, m \in (0, 1]$.

These models investigate the effectiveness of the integration policies implemented by the Philippine government. The results suggest that the integration policy which involves the creation of an environment where multiple cultural groups can coexist is the most successful. Assimilating two groups into one culture in the Modernism model was unsuccessful [102].

A population model developed by Johnson and Madin was based upon the Logistic differential equation. This model makes use of population size (N), recruitment (r), carrying capacity (K), and mortality (m) to investigate the dynamics in the insurgent population. The discrete time logistic model takes on the following form

$$\Delta N = r\left(1 - \frac{N}{K}\right)N$$
$$N_{t+1} = Nt + r\left(1 - \frac{N_t}{K}\right)N_t - m_t. \tag{2.15}$$

The model is applied to counterinsurgencies in Malaya (1948-1960) and Iraq (2003-2006) making use of data from United Kingdom Royal Air Force records and the Brookings Institution respectively. Given the available data, a least–squares optimization was implemented to estimate the unknown parameters (K, r), which are assumed to remain constant through the time period. After fitting the parameters, future trajectories were calculated using Equation 2.15. The results in the Iraq model suggested that 1) if sectarian violence had remained at low levels (such as 2006), the insurgency would have collapsed in 4-5 years based upon the US maintaining the trend of improving military performance, 2) moderate changes to the combination of K, r, or m may have led to the defeat of the insurgency in 6-12 months. Johnson and Madin suggest that increase in sectarian violence was the reason that the second case did not take place [47].

2.10 Dynamic Programming

Dynamic programming (DP) is a mathematical tool to analyze sequential decision making. Whether the problem is deterministic or stochastic, discrete or continuous

28

time, or have a finite or infinite time horizon the underlying principle problem at hand is how to sequence decisions which minimize (or maximize) some objective function. While DP takes on a wide range of problem formulations the one considered in this research is the Deterministic Continuous–Time formulation. Pertinent theory will be reviewed in this document, more complete descriptions can be found in Bellman [8], Bertsekas [11], and Denardo [33].

2.10.1   Deterministic Continuous–Time Dynamic Programming.

A continuous–time dynamic system can be described as

$$\dot{x}(t) = f(x(t), u(t)), \quad 0 \le t \le T, \tag{2.16}$$

$$x(0) : \text{given},$$

where

$x(t) \in \Re^n$	is the state vector at time t
$\dot{x}(t) \in \Re^n$	is the vector of first order derivatives at time t
$u(t) \in \Re^m$	is the control vector at time t
U	is the set of admissible controls
T	is the terminal time.

The components of $f, x, \dot{x}$, and u are denoted as $f_i, x_i, \dot{x}_i$, and u_i and the system (2.16) then represents the following first order differential equations

$$\frac{dx_i(t)}{dt} = f_i\left(x(t), u(t)\right), \qquad i = 1, \ldots, n. \tag{2.17}$$

The vectors– $\dot{x}(t), x(t),$ and $u(t)$ are column vectors and the system, f_i is assumed to be continuously differentiable with respect to x and u. The control functions (or control trajectories) are piecewise continuous functions $\{u(t) \mid t \in [0, T]\}$ with $u(t) \in U \ \forall \ t \in [0, T]$. Additionally, it is assumed that for any admissible control $\{u(t) \mid t \in [0, T]\}$, the system of differential equations (2.16) has a unique solution, $\{x^u(t) | t \in [0, T]\}$, which is the corresponding state trajectory.

The goal of this problem is to find the control trajectory $\{u(t) \mid t \in [0, T]\}$, when coupled with the state trajectory $\{x^u(t) \mid t \in [0, T]\}$, minimizes the cost function

$$h(x(T)) + \int_0^T g(x(t), u(t))dt \tag{2.18}$$

where h is the terminal cost, h and g are continuously differentiable with resect to x, and g is continuous with respect to u.

The Hamilton–Jacobi-Bellman (HJB) Equation is a partial differential equation, which under certain assumptions, is satisfied by the optimal cost–to–go function. The HJB Equation is the continuous–time analog to the DP Algorithm. The application of DP to a discrete–time approximation of a continuous–time optimal control problem is demonstrated.

First, divide the time horizon, $[0, T]$ into N pieces by

$$\delta = \frac{T}{N},$$

denote

$$x_k = x(k\delta), \qquad\qquad k = 0, 1, \ldots, N$$

$$u_k = u(k\delta), \qquad\qquad k = 0, 1, \ldots, N,$$

approximate the continuous–time system by

$$x_{k+1} = x_k + f(x_k, u_k) \cdot \delta$$

and the cost function by

$$h(x_N) + \sum_{k=0}^{N-1} g(x_k, u_k) \cdot \delta$$

Now apply DP to the discrete–time approximation. Let

$J^*(t, x)$: Optimal cost–to–go at time t and state x for the continuous–time problem

$\tilde{J}^*(t, x)$: Optimal cost–to–go at time t and state x for the discrete–time problem

The DP equations are:

$$\tilde{J}^*(N\delta, x) = h(x),$$

$$\tilde{J}^*(k\delta, x) = \min_{u \in U} \left[g(x, u) \cdot \delta + \tilde{J}^*((k+1) \cdot \delta, x + f(x, u) \cdot \delta) \right], \quad k = 0, \ldots, N-1$$

where $h(x)$ is the terminal cost. Assuming that $\tilde{J}^*$ is differentiable, it can be expanded to a first order Taylor series as such:

First define

$$F(t) = \tilde{J}^*(k\delta + \delta t, x + f(x, u) \cdot \delta t)$$

and

$$F'(t) = \left[\frac{\partial}{\partial k\delta} + \frac{\partial}{\partial x} \right] \tilde{J}^*(k\delta + \delta t, x + f(x, u) \cdot \delta t)$$

$$= \frac{\partial}{\partial k\delta} \tilde{J}^*(k\delta + \delta t, x + f(x, u) \cdot \delta t) \frac{\partial k\delta}{\partial t} + \frac{\partial}{\partial x} \tilde{J}^*(k\delta + \delta t, x + f(x, u) \cdot \delta t) \frac{\partial x}{\partial t}$$

$$= \frac{\partial}{\partial k\delta} \tilde{J}^*(k\delta + \delta t, x + f(x, u) \cdot \delta t) \cdot \delta + \frac{\partial}{\partial x} \tilde{J}^*(k\delta + \delta t, x + f(x, u)) \cdot \delta f(x, u).$$

Then

$$F(1) = F(0) + \frac{F'(0)}{1!} + \cdots + \frac{F^{(n)}(0)}{n!} + \frac{F^{(n+1)}(\theta)}{(n+1)!}$$

Since the expansion is evaluated at the point $(k\delta, x)$ this gives

$$F(0) = \tilde{J}^*(k\delta, x)$$

Next, $F'(0)$ is calculated:

$$F'(0) = \left[\delta \frac{\partial}{\partial k\delta} + \delta f(x, u) \frac{\partial}{\partial x} \right] \tilde{J}^*(k\delta + \delta t, x + f(x, u) \cdot \delta t) \Big|_{t=0}$$

$$= \left[\delta \frac{\partial}{\partial k\delta} + \delta f(x, u) \frac{\partial}{\partial x} \right] \tilde{J}^*(k\delta, x)$$

This yields

$$F'(0) = \nabla_{\delta k} \tilde{J}^*(\delta k, x) \cdot \delta + \nabla_x \tilde{J}^*(\delta k, x)' f(x, u) \cdot \delta$$

The term $\nabla_x \tilde{J}^*(\delta k, x)$ is transposed because both $\nabla_x \tilde{J}^*(\delta k, x)$ and $f(x, u)$ are vectors in $\mathbb{R}^2$ and this maps $\mathbb{R}^2 \to \mathbb{R}$. Recall that

$$F(1) = F(0) + \frac{F'(0)}{1!} + \cdots + \frac{F^{(n)}(0)}{n!} + \frac{F^{(n+1)}(\theta)}{(n+1)!}$$

Which provides:

$$\tilde{J}^*((k+1) \cdot \delta, x + f(x, u) \cdot \delta) = \tilde{J}^*(\delta k, x) + \nabla_t \tilde{J}^*(\delta k, x) \cdot \delta + \nabla_x \tilde{J}^*(\delta k, x)' f(x, u) \cdot \delta + o(\delta)$$

where $o(\delta)$ represents the second order terms satisfying $\lim_{\delta \to 0} o(\delta)/\delta = 0$. Substituting this into the DP equation:

$$\tilde{J}^*(k\delta, x) = \min_{u \in U} \left[g(x, u) \cdot \delta + \tilde{J}^*(\delta k, x) + \nabla_t \tilde{J}^*(\delta k, x) \cdot \delta + \nabla_x \tilde{J}^*(\delta k, x)' f(x, u) \cdot \delta + o(\delta) \right]$$

Since $\tilde{J}^*(\delta k, x)$ is not a function of u, subtract $\tilde{J}^*(\delta k, x)$ from both sides and rewrite the above equation as:

$$0 = \min_{u \in U} \left[g(x, u) \cdot \delta + \nabla_t \tilde{J}^*(\delta k, x) \cdot \delta + \nabla_x \tilde{J}^*(\delta k, x)' f(x, u) \cdot \delta + o(\delta) \right]$$

Then divide each term by δ and take the $lim_{\delta \to \infty}$ recalling that $\lim_{\delta \to 0} o(\delta)/\delta = 0$ gives:

$$0 = \min_{u \in U} \left[g(x, u) + \nabla_t \tilde{J}^*(\delta k, x) + \nabla_x \tilde{J}^*(\delta k, x)' f(x, u) \right]$$

assuming that the continuous equation achieves its discrete function as we take the limit; in other words:

$$\lim_{k \to \infty, \delta \to 0, \delta k = 0} \tilde{J}^*(\delta k, x) = J^*(t, x), \quad \forall\, t, x$$

The result is:

$$0 = \min_{u \in U} \left[g(x, u) + \nabla_t J^*(t, x) + \nabla_x J^*(t, x)' f(x, u) \right], \quad \forall\, t, x \qquad (2.19)$$

with the boundary condition of:

$$J^*(T, x) = h(x)$$

This is the partial differential equation known as the Hamiliton-Jacobi-Bellman equation.

Sufficiency Theorem [11]. *Suppose $V(t, x)$ is a solution to the HJB equation; that is, V is continuously differentiable in t and x, and is such that*

$$0 = \min_{u \in U}[g(x, u) + \nabla_t V(t, x) + \nabla_x V(t, x)' f(x, u)], \quad \forall\, t, x \tag{2.20}$$

$$V(T, x) = h(x) \quad \forall x \tag{2.21}$$

Suppose also that $\mu^*(t, x)$ attains the minimum in 2.20 $\forall t$ and x. Let $\{x^*(t) | t \in [0, T]\}$ be the state trajectory obtained from the given initial condition $x(0)$ when the control trajectory $u^*(t) = \mu^*(t, x^*(t))$, $t \in [0, T]$ is used [that is, $x^*(0) = x(0)$ and $\forall t \in [0, T]$, $\dot{x}^*(t) = f(x^*(t), \mu^*(t, x^*(t)))$; we assume that this differential equation has a unique solution starting at any pair (t, x) and that the control trajectory $\{\mu^*(t, x^*(t)) | t \in [0, T]\}$ is piecewise continuous as a function of t]. Then V is equal to the optimal cost–to–go function, i.e.,

$$V(t, x) = J^*(t, x), \quad \forall\, t, x.$$

Furthermore, the control trajectory $\{u^*(t) | t \in [0, T]\}$ is optimal.

Recalling Equation 2.19 the Sufficiency Theorem suggests that for an initial state, x_0, the control trajectory $\{u^*(t) | t \in [0, T]\}$, is optimal with corresponding state trajectory, $\{x^*(t) | t \in [0, T]\}$, then $\forall t \in [0, T]$

$$u^*(t) = \arg \min_{u \in U} \left[g(x^*(t), u) + \nabla_x J^*(t, x^*(t))' f(x^*(t), u) \right] \tag{2.22}$$

This is basically the minimum principle.

Minimum Principle [11]. *Let $\{u^*(t)|t \in [0,T]\}$ be an optimal control trajectory and let $\{x^*(t)|t \in [0,T]\}$ be the corresponding state trajectory, i.e.,*

$$\dot{x}(t) = f(x^*(t), u^*(t)), \qquad x^*(0) = x(0) \ : \ given$$

Let also $p(t)$ be the solution of the adjoint equation

$$\dot{p} = -\nabla_x H(x^*(t), u^*(t), p(t)),$$

with the boundary condition

$$p(T) = \nabla h(x^*(T)),$$

where $h(\cdot)$ is the terminal cost function. Then, for all $t \in [0,T]$,

$$u^*(t) = \arg\min_{u \in U} H(x^*(t), u, p(t)).$$

Furthermore, there is a a constant C such that

$$H(x^*(t), u, p(t)) = C, \qquad \forall t \in [0,T].$$

2.10.2 Approximate Dynamic Programming.

With a continuous-time dynamic system based upon a nonlinear piece-wise differential equation such as

$$\dot{x}_i(t) = \begin{cases} 0, & \text{if } x_i(t) + \frac{\partial x_i}{\partial t} \leq 0; \\ f_i\left(x_i(t), u(t)\right), & \text{if } 0 < x_i(t) + \frac{\partial x_i}{\partial t} < 1; \\ 1, & \text{if } x_i(t) + \frac{\partial x_i}{\partial t} \geq 1. \end{cases} \qquad (2.23)$$

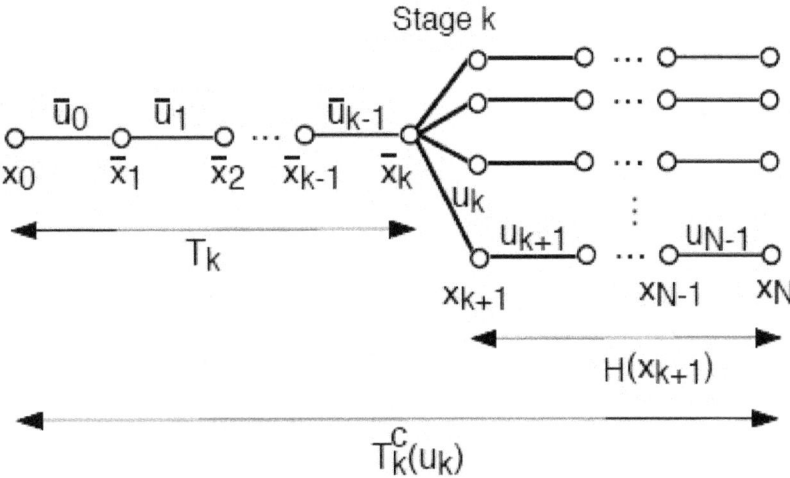

Figure 3. Graphical depiction of a rollout algorithm [12]

various problems with determining closed–form solutions arise. In cases like this, the HJB equation does not apply. The field of approximate dynamic programming addresses these types of problems. While it does not provide an true optimal solution it provides near–optimal solutions to problems where an exact solution does not exist or cannot be determined. One such type of technique is called rollout algorithms which is a suboptimal control method that relies on a suboptimal policy, a base heuristic.

2.10.3 Rollout Algorithms.

A rollout policy is a one–step lookahead policy with the optimal cost–to-go approximated by the cost–to–go of the base policy $(\tilde{J}_{k+1}^*)$ [11]. The basic formulation of a rollout algorithm can be observed graphically in Figure 3. In Figure 3 at stage k, given the current partial trajectory (T_k) which starts at x_0 and ends at $\bar{x}_k$, the rollout algorithm evaluates all possible state transitions $x_{k+1} = f_k(\bar{x}_k, u_k), u_k \in U_k(\bar{x}_k)$, and runs the base heuristic starting with x_{k+1}. Then the approach finds a control

$\bar{u}_k \in U_k(\bar{x}_k) \ni T_k^c(\bar{u}_k) = T_k \cup (\bar{u}_k) \cup H(\bar{x}_{k+1})$ where $\bar{x}_{k+1} = f_k(\bar{x}_k, \bar{u}_k)$, is feasible and has a minimum cost.

Rollout algorithms for approximate solutions of discrete optimization problems are shown in Bertsekas and Tsitsiklis [14]. Bertsekas, Tsitsiklis and Wu [15] provide rollout algorithms to address combinatorial optimization which improve the performance of the base heuristic. Bertsekas and Castañon [13] developed rollout algorithms to find near–optimal solutions to stochastic scheduling problems with considerable savings in computation time over the base heuristic. In 2005 Bertsekas [16] applies rollout algorithms to constrained deterministic programming problems. A 2005 and 2013 survey of Approximate Dynamic Programming and rollout algorithms provide rollout algorithms for a wide range of discrete optimization problems [12], [17]. Applications of rollout algorithms are found in many fields to include: logistics [10] and [103]; vehicle routing [86], [67], [43] and [90]; and sensor scheduling [48] and [62]. Rollout algorithms are not presented in any literature for use in nation–building problems.

2.11 Summary

In this chapter, a review of relevant background literature is presented to provide the context for the creation of the system of differential equations model and the application of dynamic programming in this field. The two key points from this section are 1) the concept of the unilateral prescriptive model using the DIME-PMESII paradigm and 2) the concept of using dynamic programming to solve these nation-building problems.

After reviewing the literature there is a need for a prescriptive model which considers the tools (DIME) that influence key macro level variables (PMESII). No model addresses this need; several models consider populations (Lanchester type attrition models) and some integrate external forces. Few models address more than one aspect of DIME. The ones that do use it as a measure to determine force levels. There is a distinct gap in literature concerning prescriptive models which consider a robust set of tools the US can implement in nation building operations.

Additionally, the use of rollout algorithms in optimization problems is demonstrated in the literature yet not in the social science fields, particulary in the context of nation–building problems.

The model and algorithms presented in this dissertation specifically addresses these gaps.

III. Investigating the Dynamics of Nation Building Through a System of Differential Equations

3.1 Abstract

Nation–building modeling is an important field given the increasing number of candidate nations and the limited resources available. In this research we present a modeling methodology and a system of differential equations model to investigate the dynamics of nation–building. The methodology is based upon solving inverse problems, much like Lanchester Equations, and provides Measures of Merit (MoM) to evaluate nation–building operations. An application is derived for Operation Iraqi Freedom (OIF) to demonstrate the utility as well as effects of various alternate strategies, using differing applications of national power. This modeling approach is data driven and offers a significant, novel capability when analyzing and planning for future nation–building scenarios.

3.2 Introduction

The US has and continues to aid and assist those nations, when needed, in order to prevent them from becoming safe–havens for terrorist and extremist activity and develop a sustainable and viable peace. The use of an armed force and economic aid to promote political and economic reform with the objective to assist a nation in transition from conflict to peace (internally and bordering nations) is nation–building [38]. These are not unilateral military or State Department operations, but rather the synchronous effort of military and civilian, public and private, as well as US and international efforts to provide assistance to the state or region in need [88]. The National Security Presidential Directive 44 [23] and the National

Security Strategy [69] state a key interest of the United States are those states which are in transition or reconstruction.

According to Joint Publication (JP) 3-0, *Joint Operations* [35], there are four instruments of national power that can be applied by the US, these represent exogenous actions on an operational environment; they are, diplomacy, informational, military, and economic (DIME). Within the operational environment of a nation the Army Field Manual (FM) 3-0, *Operations* [36] describes six endogenous and interrelated variables that describe its internal state and give insight to its progress; they are, political, military, economic, social, information, and infrastructure (PMESII). This is often referred to as the PMESII-DIME paradigm, where the effects of the DIME actions are measured through PMESII.

Leading military operations analysts have conjectured that a system of differential equations could be developed that demonstrate the effect of resources in nation–building [21]. These variables, external and internal, form the basis of the nation–building model described in this research. This problem takes the form of an inverse problem. In an inverse problem, the results or effects of the problem are known and measurable, but the model's underlying structure, the cause, is uncertain [45]. Thus, the nature of such problems is to resolve cause from effect. There exist models which look at operational effectiveness, models which predict instability [87], models that determine which countries are candidates to provide nation–building [59], and models which determine a framework for scheduling reconstruction operations [27]. However, there are no models which look at the tools available to conduct nation–building and evaluate the impact through the variables which describe the operational environment as indicated by [87] and [31]. A model, based on classical Lanchester models, would be very useful "to describe effects of inputs and interactions of state variables." [21]

This research develops a nation–building model through a solution methodology to the inverse problem. A representative system of differential equations, using the nation–building operations in Iraq as a framework is presented. Additionally, this research shows how the endogenous operational variables (PMESII) can act as Measures of Merit (MoM), against which we can evaluate different applications of national power (DIME).

The next section is a review of literature of the PMESII-DIME paradigm and military applications of differential equations to inverse problems. In section three, we describe our solution methodology by outlining our data; organizing the data into different composite indices and MoM; then solve the system of differential equations. Section four offers a description of the model to evaluate how our MoM are changed when alternate strategies are undertaken by the US, using data from Operation Iraqi Freedom (OIF). Finally, conclusions and future research are presented in the final section.

3.3 PMESII-DIME Paradigm

The concept of the PMESII-DIME modeling paradigm has been prominent for several years. With the beginning of the "Global War on Terror" the need for these type of models increased as the US military and its coalition partners quickly overmatched regimes and then became involved in lasting counterinsurgency and nation–building operations.

The Measuring Progress in Conflict Environments (MPICE) project in the mid–2000s was important in several ways. First, while it did not make direct use of PMESII, it did consider the political, security, rule of law, economic, and social aspects of a country. Secondly, it addressed a gap in the current operations and policy. This gap being a lack of metrics which assist in formulating policy and implementing operational and strategic plans to transform nations at risk or in conflict, and bring stability to war-torn societies [39]. This identified a need for tools to address nation–building operations; PMESII-DIME models could address this need.

Over the years there have been several tools which fit into this framework. A partial list is provided by [44]. More recent approaches used to model PMESII-DIME include a wargaming through the Peace Support Operation's Model (PSOM) [20], agent based tools such as Senturion [2], the Power Structure Toolkit (PTSK) [94], and the intelligent agent approach demonstrated by [60]. Network models such as the DynNetSim tool [3] and Polyscheme [25] can represent the multi-nodal and connected aspect of PMESII-DIME. Bayesian networks are also popular tools that have been used by [65], [75], and [58]. Other models combine several models to simulate and influence behaviors such as the Conflict Modeling, Planning, and Outcome Exploration system (COMPOEX) [53]. These models all make use of the PMESII-DIME paradigm outlined in JP 3-0 and FM 3-0.

3.4 Differential Equation Models

Within military applications, one of the most famous treatments of inverse problems are Lanchester Equations. They have been the basis for many differential equation models involving combat, however they have also been documented in other areas, such as irregular warfare. [84], for example, used Lanchester Equations to model guerilla warfare and asymmetric engagements while employing and operationalizing an array of variables and coefficients (representing weapons strength, discipline, morale, etc.) to model an insurgent force in Phase II (strategic stalemate) of an insurgency. [78] used a system of differential equations to model the arms race and instability of nation states based upon the current levels of its neighboring and/or *menacing* states.

42

Many Lanchester equations involving several parameters and inputs can be formulated as a system of differential equations. [18] model insurgency using this method and parameters based upon the recruitment and effectiveness rates of coalition and insurgent forces in OIF. [47] also derive a model which investigates the dynamics of the insurgent population, this model makes use of the logistic differential equation and the idea that there is carrying capacity involved. The model is then applied to the insurgency in Malaya (1948-1960) and the first three years of OIF (2003-2006). Both models make use of data from the Brookings Institution, *Iraq Index* [70]. In 2009 [1] provided the Formal Stabilization and Reconstruction Operations Model (FSROM) using a system of differential equations and seeming unrelated regression estimation (SURE) to determine the optimal degree of foreign aid, multilateralism (number of nations participating), and operation length. The model was tested using instances of nation–building operations from post World War II through operations in Afghanistan and Iraq with data from the [96]. In 2011 [5] present a Lanchester model to study armed revolts, where success is largely determined by the population instead of the initial force size.

The evolution of Lanchester Equations, from air combat to insurgent conflict to political and economic problems, demonstrate how the application of inverse problems to warfare has come to model more complex ideas over time. The increasingly complex and uncertain nature of warfare naturally lends itself to an inverse problem application. We incorporate the political, military, economic, social, and infrastructure (PMESI) variables of the operational environment and the military and economic (ME) instruments of national power . We rely on collected historical data from established institutions (e.g. [96], Brookings Institute [70]) to inform our model. The additional PMESII-DIME variables were excluded due to a lack of available data.

3.5 Solution Methodology

The developed solution methodology to solve the inverse problem involves three steps, depicted in Figure 4.

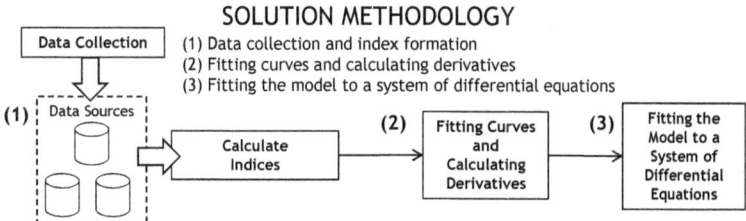

Figure 4. Solution Methodology

Each step is described in the following sections, and later applied to OIF in Section 3.6.

3.5.1 Data Collection and Index Formation.

The PMESI indices are formulated as composite indices. They represent a mathematical transformation (and aggregation) of different relevant indicators into one value. The use of such indicators to reflect country performance is widely practiced by several organizations (i.e. [96]). A survey by [6] details over 170 different composite country performance measurements used in practice. The *Handbook on Constructing Composite Indicators* [66], published by the Organisation for Economic Co-operations and Development, notes that several key attributes (summarize complex realities, reduce indicators, assess nations over time, and compare complex dimensions) that composite indices accomplish. Nardo *et al.* warn that the justification and construction of composite indices lies in their fitness to the intended purpose and the acceptance of peers. Following this, we build our composite indices following the PMESI operational variables outlined in the Army Field Manual (FM) 3-0, *Opera-*

tions, Joint Publication (JP) 3-0, *Joint Operations* and the literature referenced in Section 3.3.

Define $X \in \{P, M, E, S, I\}$ as the set of operational variables. Let $X^{(j)}$ represent the *jth* component of X. Each $X^{(j)}$ is composed of a set of components or indicators, i, with each of the i indicators containing n data points. We can see that $n \mapsto i$, $i \mapsto X^{(j)} \Leftrightarrow n \mapsto X^{(j)}$ where particular indicators are used to calculate one and only one of the PMESI variables, $X^{(j)}$, and that $X^{(j)}$ represents the composite index at a time, t; we will refer to this as $X_t^{(j)}$.

The indicators take measurements from the same operational environment, or space; however, each of the data will vary in unit and range. To account for this, a normalization technique is applied to place each indicator on a common $[0, 1]$ scale. To use the normalization we establish benchmark values, a best and worst value; this is the maximum and minimum observation value over the entire range of observations for each component. This ensures that all values will assume a normalized score on a common scale. The result is that i_t assumes a normalized value according to the established benchmarks and the raw score of the indicator. This allows for a common comparison and allows for the weighting of the indicators to calculate and overall index score, $X_t^{(j)}$.

Each indicator then can be assigned a weight, w_i, which indicates the percentage of an indicator to an index; the total weights must not exceed 1. The weighting can be accomplished through various methods based on the preference of the analyst and decision maker. The result is a set of index vectors organized by time. In this application the weights are determined through the rank ordered centroid method based on subject matter expert (SME) opinion.

3.5.2 Curve Fitting and Calculating Derivatives.

A consideration in building these complex models is that the nature of the data is complex and often difficult to interpret as well. The use of curve fitting and smoothing is an available technique to observe the trend of data which may not be readily apparent from the data itself. The use of this technique was demonstrated with a complex composite index, the Dow Jones Industrial Index, by [49] to observe trends while still considering the volatility of the market.

Due to the error injected into the data by measurement error and other sources, we *smooth* the normalized data by fitting curves to each composite index. First, we do this to capture the general trend of the data while still accounting for potential noise. Thus, the generalized model, while not susceptible to extraordinary events, can still account for them. Second, we calculate the derivatives of our fitted curve at each point, t, and use them to approximate the derivative function of our operational variables. This is an integral part in calculating the coefficients of the final system of differential equations, completed in the third step.

The curve fitting process can be accomplished through various methods; however, the method selected must pass an appropriate goodness–of–fit test based on the operational situation and must place emphasis on matching the end effects. Due to the curvilinearity of the data, a weighted polynomial regression is selected. Once a curve has been fit, we have a general equation that will approximate the index values. From this equation a derivative can be calculated which can then be used to approximate a derivative for each t.

Through the first two steps of this process, observations of data over time are collected and compiled into a composite index, which summarizes complex data and assesses the progress over time as described by [66]. This develops a set of discrete points which describe the current state of the PMESI variables at a time, t. The general trend of the data helps to describe its progress; to capture this trend a regression technique is used. After selecting the proper fit, the PMESI indices can be expressed as a function of time. The derivative of the function is then calculated and used to determine the coefficients of the system of differential equations in the least–squares minimization. Through this series of steps, mean–field theory was applied to perform the necessary steps to determine the coefficients of the system of differential equations in the next section.

3.5.3 Determine the Coefficients to the System of Differential Equations.

The model must encompass the interactions within the operational environment and the corresponding endogenous variables. To introduce the interrelatedness we conjecture a system of differential equations by setting the rate of change of each PMESI variable equal to a function of the ME and PMESI factors. In its most general form we have

$$\dot{x}_t = p(P, M, E, S, I) + d(Mil, Eco) \tag{3.1}$$

where each PMESI derivative is a function of the PMESI indices and the ME forcing functions. This ensures that the interconnected systems perspective described in JP 3-0, *Joint Operations* is incorporated and will facilitate the understanding of the continuous and complex interactions within this dynamic system.

$$
\begin{aligned}
f_i(x(t), u(t)) = \;& a_{i1}\left(\frac{P_t}{b_{i1}} - 1\right) + a_{i2}\left(\frac{M_t}{b_{i2}} - 1\right) + a_{i3}\left(\frac{E_t}{b_{i3}} - 1\right) + a_{i4}\left(\frac{S_t}{b_{i4}} - 1\right) \\
& + a_{i5}\left(\frac{I_t}{b_{i5}} - 1\right) + d_{i1}Mil_t + d_{i2}Eco_t
\end{aligned}
\tag{3.2}
$$

To build the functional form of the model (Equation 3.2), we introduced three parameters, a, b, and d to the general equation. The a and d coefficients are the scaling factors, representing the weight of the endogenous functions or the proportionality of the endogenous function to the rate of change for a PMESI variable. The b coefficient is the tipping point, it represents the point where a change in the parameter causes a change in the dynamical property of the system, much like a bifurcation. The tipping point represents a point when $x_t \approx b$ the effect of a variable is generally stable, when the value of $x_t > b$ there is a magnifying effect, and when $x_t < b$ there is a diminishing effect. The d parameter is the scaling factor coefficients for the Mil and Eco forcing functions. The range for a_{ij}, b_{ij}, and d_{ik} are

$$a_{ij} \in \mathbb{R} \qquad\qquad \text{for i,j=1,2,\ldots,5}$$

$$b_{ij} \in \{\mathbb{R} \mid 0 < b_{ij} \leq 1\} \qquad\qquad \text{for i,j=1,2,\ldots,5}$$

$$d_{ik} \in \mathbb{R} \qquad\qquad \text{for i=1,2,\ldots,5, for k=1,2}$$

The full system of equations can be formed by creating an equation for each $X \in \{P, M, E, S, I\}$ which corresponds to the i; j is the index within each derivative, and k corresponds to the forcing function.

This tipping point highlights the advantage of data driven model. Rather than looking for a subjective assessment from a subject matter expert (SME) the observed data determines the tipping point within the associated PMESI variable range, with range $[0, 1]$ By letting the data drive the value of the coefficients, the effect relative to the time period and the interrelatedness of the data can be evaluated. This means when evaluating the military of a country that is building its strength, the evaluation may be based upon the observations which demonstrate the development of the military, and how they interact with the other variables in the operational environment.

The system of differential equations accurately describes the actual trends of our operational variables, thus the system is fit to the derivatives of the fitted–curves from the previous section. Many problems involve a large number of independent variables where the exact calculation of such a problem is infeasible. In order to solve these problems, efficient approximation techniques are needed in order to better understand their dynamics. The method of using a Mean–Field Equation (MFE) to approximate these dynamics is an efficient approximation method to aid in solving problems dealing with uncertainty and complexity [72]. This method has been used by [51] to capture similar mean–field interactions. In order to solve for the a, b and d coefficients a nonlinear least–squares minimization problem is utilized. The b coefficients are restricted to the same range of the indices. The state indices are restricted to the range $[0, 1]$, thus the tipping points (b coefficients) must also be in the range $[0, 1]$. To maintain indices within the prescribed range the system transitions according to the following piecewise differential equation

$$\dot{x}_i(x(t), u(t)) = \begin{cases} 0, & \text{if } x_i(t) + f_i(x(t), u(t)) \leq 0; \\ f_i(x(t), u(t)), & \text{if } 0 < x_i(t) + f_i(x(t), u(t)) < 1; \\ 0, & \text{if } x_i(t) + f_i(x(t), u(t)) \geq 1. \end{cases} \qquad (3.3)$$

The minimization problem is solved using the Generalized Reduced Gradient (GRG) method for solving nonlinear programs. There are two points to note in this case:

1. Just like many nonlinear problems, the solution for a_{ij}, b_{ij}, and d_{ik} may not be unique. It is typical that if there is more than one solution, then there is an infinite number of solutions that satisfy the equations.

2. The solution provided is specific to the operational environment being studied. There is no master set of coefficients or parameters that can be used for all situations. The so–called "constant fallacy" described by [45] is often overlooked

and leads researchers to believe they have found universal parameters when they have in fact found parameters specific only to their study.

The results provide a mathematical expression of the operational environment, but with far more insight and capability than the original fitted curves from the previous section. Using the solved system of differential equations, modifications to the instruments of national power used by the US in terms of military troops and economic aid can be explored to see how these changes effect the evolution of the state variables for the nation of interest undergoing nation-building while capturing the interactions between the operational variables and the impact of external influences.

In the next section, the methodology described in Section 3.5 is implemented using a data set from OIF. The implementation method and results, as well as analysis from alternate ME strategies, are presented in Section 3.6.

3.6   Implementation

This section illustrates the utility of the model through an application of data from OIF.

Data was collected to construct each of the PMESI variables starting with the beginning of the war (March 2003) through December 2008 from the following sources: Brookings Institution, Department of Defense, Department of State, and the CIA Factbook. Each component was normalized, a notional weighting scheme developed, and index values were calculated for each value, t. Each individual index is then plotted as a time series. A 4th order polynomial expression is the result of the weighted regression step. The 4th order polynomial was selected as it was the lowest order polynomial with all indices having at $R^2 > .8$ with matching end effects. The derivatives of the 4th order polynomial equations at each point, t, are used to approximate the derivative function of the operational variables.

To fit the system of differential equations, the coefficients are solved for using the nonlinear least–squares method. Using the nonlinear program, the values of a_{ij}, b_{ij}, and d_{ik} coefficients are calculated which minimize the SSE. The a, b, and d coefficients (truncated values shown here) are provided in Tables 1–2.

Table 1. a **and** b **coefficients**

i ＼ j	Political	Military	Economic	Social	Infrastructure
Political (a)	-0.0108181	-0.0200317	0.0000199	0.0040096	0.0035656
Military (a)	0.0743136	0.0168815	0.0002164	-0.0101848	-0.0241455
Economic (a)	0.0287454	0.0155291	-0.0008631	-0.0100445	-0.0156717
Social (a)	-0.0005479	0.0001096	0.0004305	0.0246105	0.0021363
Infrastructure (a)	0.0034676	0.0008256	-0.0010147	-0.0264765	-0.0054859
Political (b)	0.3980158	0.6319466	0.0061526	0.0894118	0.1010944
Military (b)	0.9093994	0.5026149	0.5234950	0.1443637	0.6481461
Economic (b)	0.4439128	0.3606699	0.1873238	0.1093262	0.3084922
Social (b)	0.1788101	0.0781656	0.0700705	0.4343948	0.2960607
Infrastructure (b)	0.2387622	0.0769009	0.1342082	0.3479762	0.2603047

Table 2. d **coefficients**

i ＼ k	Military	Economic
Political	-0.0074973	-0.0008035
Military	0.0230743	0.0007235
Economic	0.0184431	0.0007638
Social	-0.0017058	0.0008385
Infrastructure	0.0049494	-0.0005529

The calculated derivatives indicate the rate of change in the system for each operational variable. The derivatives provide useful information; however, if the rate of change and a starting point are known, then a calculated index $(\hat{P}, \hat{M}, \hat{E}, \hat{S}, \hat{I})$ can be used to gain more insight. This is an initial value problem. One method of solving first-order differential equations with a numerical method is the Euler method, which uses the derivative and the initial value to estimate the solution ($u_{k+1} = u_k + ha_k$ for $k = 0, 1, \ldots, n$ where $a_k = u'_k$ and h is the step size [42]). Through an ap-

51

plication of this method, the index values can be estimated through the knowledge of an initial value and the derivatives provided from the system of differential equations. The forward Euler method is unstable if any of the eigenvalues (λ) have 0 or positive real parts. In linear form the model can be expressed as $Ax + Bu + c$ where A is a matrix of $\frac{a_{ij}}{b_{ij}}$, B is the matrix of d_{ik} coefficients and c is the vector containing the the sum of a_{ij} for each i. Although $\lambda = [0.0414, -0.0019 - 0.0361i, -0.0019 + 0.0361i, -0.002, 0]$ and indicates that $\frac{\partial}{\partial t} \rightarrow \pm\infty$ for some values; $\dot{X}_i(t)$ is truncated to values between 0 and 1 according to Equation 3.3. Using this method, with $h = 1$, the approximate index is calculated and compared to the original index values in Figure 5.

An Anderson-Darling Goodness-of-Fit test (Table 3) was conducted using [64]. The results indicate that the Political, Military, and Infrastructure error are normally distributed, indicating the successful application of the mean-field approximation.

Table 3. Anderson-Darling Goodness of Fit

$\alpha = 0.05$	
Political	0.560
Military	0.427
Economic	< 0.05
Social	< 0.05
Infrastructure	0.145

Given the quality of the data and the nature of the problem this is not entirely unexpected. This indicates the mean-field method captures the general trend of the variable it is approximating when the data that provides the approximation is appropriate. The inference here is that the data which builds the Economic and Social indices may be biased, collection methods may have changed, or the data needs to be improved, indicating an area for possible future research.

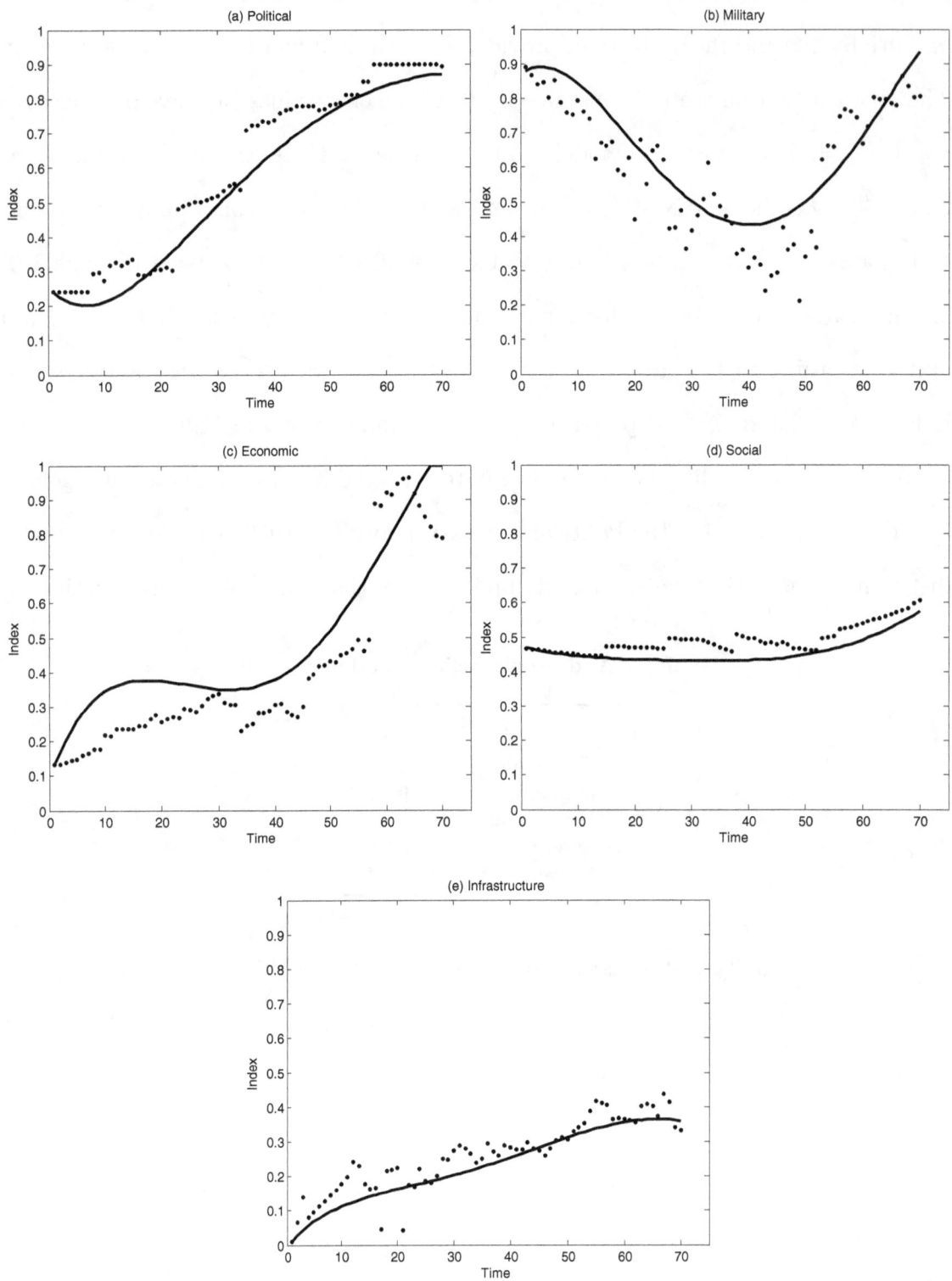

Figure 5. Calculated Indices versus Actual Indices– The solid lines represent the calculated indices from the model using the Euler method and the points are the actual monthly index values.

3.7 What–If Analysis

In this section the methodology is applied to evaluate two alterative strategies which reflect possible modifications to the military influence as applied by the US in Iraq. The following sections organize modifications to economic and military data (on the side of the US) into different strategies. These modifications are derived from actual implemented plans, proposed legislation from Congress, and demonstrate the what-if analysis feature of the model.

3.7.1 The No Surge Alternative.

On the 10th of January 2007, President George W. Bush delivered a speech to the American Public outlining a new strategy in Iraq. As part of that strategy he called for the additional deployment of 20, 000 US troops, the majority deploying to Baghdad to help Iraqis clear and secure neighborhoods, help them protect the population, and help ensure that the Iraqi forces left behind are capable of providing the security needed [24]. The first deployment of troops was in January 2007 and in July 2007 all surge troops had been deployed. The surge would last to July 2008 and was roughly an increase of 28, 000 troops [70].

The actual US troop numbers for the Mil_t variable represent the surge strategy and serve as the base case for evaluating the alternate strategies. In order to evaluate the no surge strategy, the Mil_t are adjusted under the assumption that if the surge was not implemented, the number of troops would have remained the same during for the time period. Therefore, the number of troops are held constant from the January 2007 level through February 2009 (when the troop level returned to near the pre–surge level). All other variables remain the same, specifically the coefficients are not changed as the goal is to evaluate the alternative strategy under the conditions that took place. In other words, if everything else remained the same how would the indices have been affected by the no surge policy? The results are shown in Figure 6. The observation here is how the indices change over time based on a different troop level, while keeping all other variables constant. The change in Mil_t variables did have impacts on all PMESI index variables, some more than others. The plots are identical up until month 48 when the surge began, and then takes a different trajectory based upon the changes in the Mil_t variables and the corresponding interactions from the PMESI index variables. As a result, the trajectory of the Political and Social indices did not differ greatly from the original values over time while the Military, Economic, and Infrastructure trajectories decreased over time. 95% confidence intervals were calculated at the end of the 70 month time period that show that these differences, given this data and model, would not be statistically significant.

3.7.2   A Complete Reduction by 2008 Alternative.

In March–July of 2007, Congress proposed a series of resolutions that would lead to the removal of US troops in Iraq. The first one, House Resolution (H.R.) 1951 was passed by Congress and vetoed by President Bush; H.R. 2956 was passed by the House and required the Secretary of Defense to initiate the reduction of troops in Iraq immediately through April 1, 2008. This resolution was then sent to the Senate where it was narrowly defeated 52-47. This resolution was never introduced again and the current surge plan continued as outlined in January of 2007 [95].

To evaluate the potential impact of the withdrawal strategy, we assume that the March Resolution passed and the number of troops were reduced in even increments over the next year leading to no troops in April 2008. All other variables remain the same, as in the previous case. In other words, if everything else remained the same how would a phased withdrawal implemented in 2007 affect the indices? The results are shown in Figure 7.

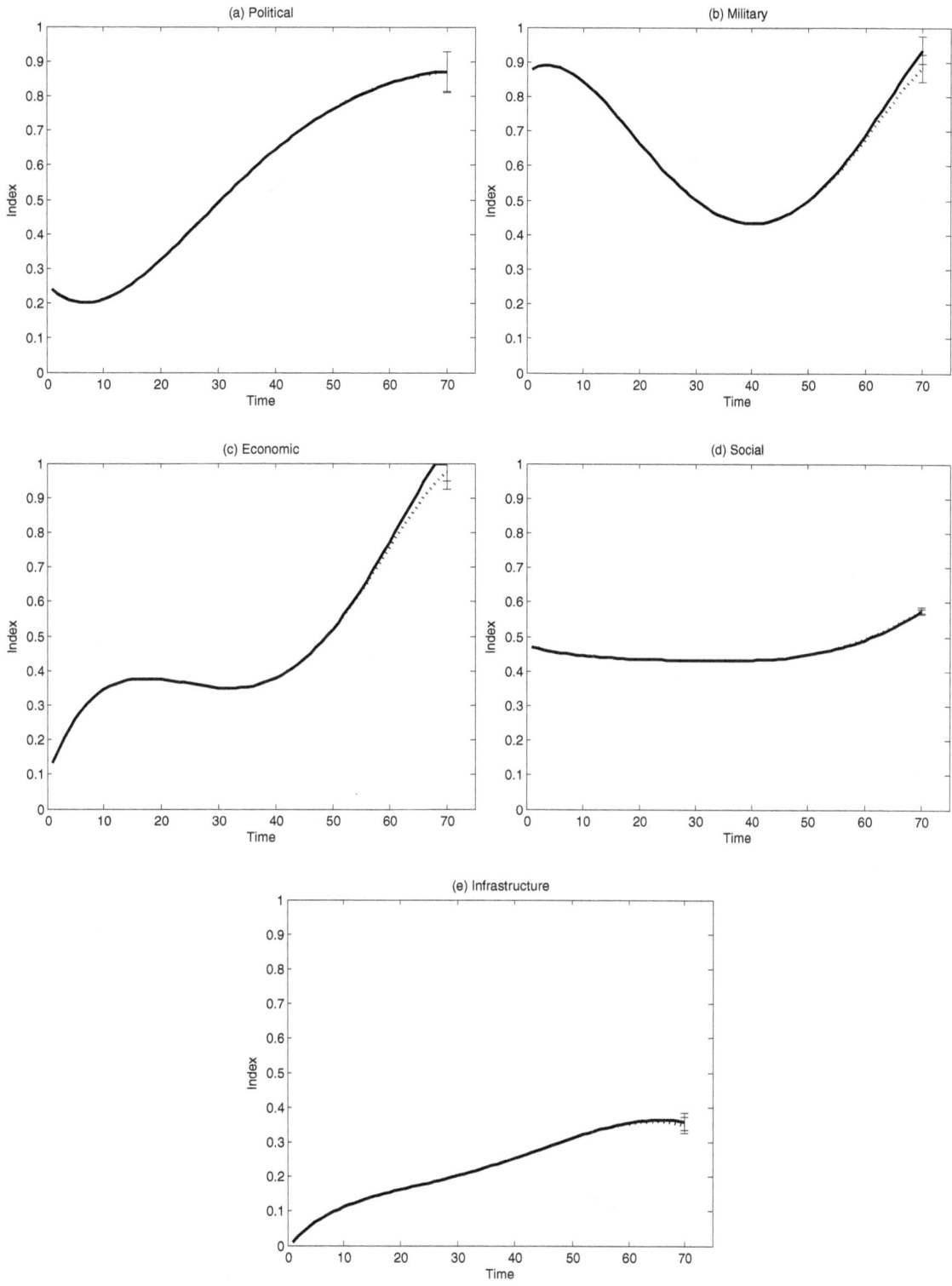

Figure 6. No Surge (dashed line) vs Surge (solid line)– The calculated index plots from the model using the Euler method to compare the alternative strategy (No Surge) to the actual strategy (Surge).

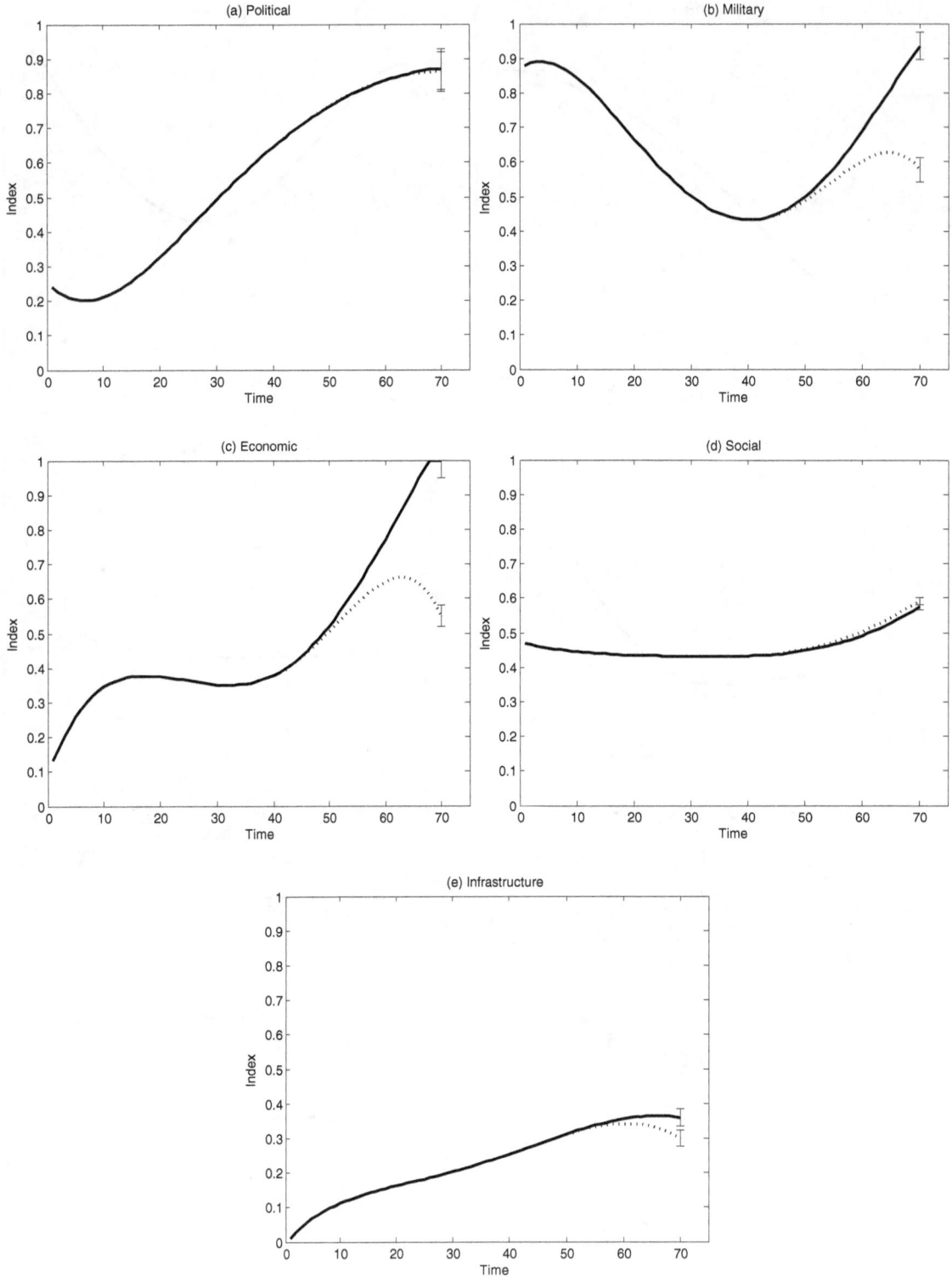

Figure 7. Withdrawal (dashed line) vs No Withdrawal (solid line)– The calculated index plots from the model using the Euler method to compare the alternative strategy (Withdrawal) to the actual strategy (No Withdrawal).

The observation here is how the indices changed over time based upon the withdrawal troop level and timeline, with all other variables constant. The change in the Mil_t variables did have effects on all the variables, some more than others. The plots are exactly identical up until month 50 where the withdrawal began and then takes on a different trajectory, based upon the interactions from the different data. As a result, the trajectory of all indices except Social decreased over time, with significant deviations from the actual plot in the Military, Economic, and Infrastructure variables. After interpreting the results it can be seen that the withdraw policy had projected impacts on all of the PMESI indices.

As conjectured, the model provides a means to investigate various strategies through changing the inputs into the model. Changing an exogenous variable may result in a difference which can have an increasing or decreasing effect. One can observe that there are rewards and costs associated with changing the inputs dependent on the interrelated dynamics. Once again, it is important to note this set of coefficients and equations are based on the data used to build the model.

3.8   Conclusion

In support of efforts to develop analytical methods for use in Irregular Warfare operations, this research develops a methodology that addresses nation–building and accounts for the impacts of the instruments of national power. The developed model captures the interrelatedness and complexities reflective of an actual operational environment.

The shift in warfare as described in FM 3-0 has gone from *around the people* to *among the people* [36]. This change marks a new paradigm beyond how we fight and into how we plan our operations. The ability to measure the PMESII variables parallels the work performed by military planners. If operations are planned in this context, it makes sense to measure and evaluate them in the same context. Developed models provide insight to analysts and decision makers on the application of the instruments of national power in terms of the operational environment.

The developed model methodology is data driven and offers a significant, novel approach that allows wargaming, analyzing, and planning future nation–building operations. The methodology provides the ability to inform strategic resource allocation decisions during ongoing nation–building operations. Historical examples may be modeled and analyzed using this methodology to develop an integrated comprehensive approach to future nation–building.

IV. An Approximate Dynamic Programming Approach to Resource Allocation for Nation–Building Problems

4.1 Abstract

The challenges of nation–building are faced by governments when assisting failing or failed states. At the base of this challenge exists a resource allocation problem; how to apply limited resources in a manner to maximize measurable outcomes or minimize costs. Treating the nation–building problem as a resource allocation problem requires several operations research and mathematical techniques. An approximate dynamic programming (ADP) formulation and techniques are developed to address this problem and are implemented using a system of differential equations model of the nation–building operations in Iraq to evaluate the allocation of resources. Multiple cost functions and base heuristics are presented to develop significantly improved policies for given objective functions.

4.2 Introduction

One modern approach to nation building is defined as

> the use of an armed force as part of a broader effort to promote political and economic reforms with the objective of transforming a society emerging from conflict into one at peace with itself and its neighbors [38].

This is not a new concept; in the history of the United States (US) alone there are 19 armed conflicts which can be categorized as nation building operations [73]. While this problem is not as well defined as classic optimization problems, this research will show how approximate dynamic programming (ADP), specifically rollout algorithms, can address the problem of nation–building. For this dynamic programming approach a system of differential equations model that captures the effect of external influences on the rate of change of a state of a nation is used. Significant improvement in the objective function value is achieved for several objective functions.

The literature addresses modeling the nation-building problem (or a sub-set such as counterinsurgency operations) through several different methods such as systems dynamics models, Lanchester equation and differential equation models. Richardson [77] created a systems dynamics model to capture post-reconstruction operations at a national level which Robbins [79] then modified for regional analysis. Pierson [76] developed the famous "spaghetti" diagram of the counterinsurgency effort in Afghanistan and then Minami and Kucik [63] applied a similar effort to Iraq using systems dynamics models. In 2008, Blank *et al* [18], developed a dynamic model of insurgency using Lanchester equations and Iraq war data. The model proposes a system of differential equations, the general solution of which is then used to plot the phase portraits of the system and deduce information. Johnson and Madin [47] developed a population model based upon the Logistic differential equation. This model makes use of population size, recruitment, carrying capacity, and mortality to investigate the dynamics in the insurgent population.

In 2009, Abdollahian *et al.* [1] provided the Formal Stabilization and Reconstruction Operations Model (FSROM) using a system of differential equations and Seeming Unrelated Regression Estimation (SURE) determine the optimal degree of foreign aid, multilateralism (number of nations participating), and operation length. Tauer, Nagi, and Sudit [93] formulated a simplified version of the model by Richardson [77] as a Markov decision process and used a Reduced Approximate Linear Program ($RALPH_E^H$) for the H-neighborhood around an initial state x_0, and assuming a given expert's policy π_E. This was used to model the transition between population classes (Unemployed, Private, Government).

Arney and Arney [4] use a large scale system of differential equations and a network structure to model counter-insurgency and coalition operations in stages. The network model describes the collaboration link between nodes while the system of differential equations provides the metrics to evaluate operations. Saie and Ahner [81] propose a model system of differential equations model to address the nation–building problem using a paradigm based on military planning variables. An updated version of this model provides the use case for this research's resource allocation.

Rollout algorithms are a type of ADP technique that is often used to solve complex problems that have no closed form solution. This occurs when the state space and/or control space is large and exceeds the computational capacity of computers. Rollout algorithms use a heuristic to approximate the future cost or reward of a current decision. Rollout algorithms for approximate solutions of discrete optimization problems are shown in Bertsekas and Tsitsiklis [14]. Bertsekas, Tsitsiklis and Wu [15] provide rollout algorithms to address combinatorial optimization which improve the performance of the base heuristic. Bertsekas and Castañon [13] developed rollout algorithms to find near–optimal solutions to stochastic scheduling problems with considerable savings in computation time over the base heuristic. In 2005 Bertsekas [16] applies rollout algorithms to constrained deterministic programming problems. A 2005 [12] and 2013 [17] survey of Approximate Dynamic Programming and rollout algorithms provide rollout algorithms for a wide range of discrete optimization problems. Applications of rollout algorithms are found in many fields to include: logistics [10] and [103]; resource allocation [52], [101] [28], and [61]; vehicle routing [86], [67], [43] and [90]; and sensor scheduling [48] and [62]. Rollout algorithms are not presented in any literature for use in nation–building problems.

4.3 Dynamic Programming Formulation

Our formulation addresses the resource allocation problem as applied to nation–building operations in order to improve the state of the country. A system of differential equations represents the dynamics of a nations state, to include the effect of the resources allocated. Using the differential equation, a per period and total amount of resources are applied using a specified objective function to achieve maximum improvement. This problem is one the US has faced for the past decade in Afghanistan and Iraq. This is not a unique problem to the US, as recently as 2013 the United Nations (UN) and France were compelled to intervene in Mali to restore peace and conduct nation–building operations.

Consider a country that is unstable and the goal is to employ assets to provide support to that nation. The total amount of resources and when they are allocated are critical questions to be addressed. We will adopt the PMESI-ME (PMESI–political, military, economic, social, and infrastructure; ME– military and economic) paradigm described in [81] and add the diplomatic instrument yielding PMESI-DME.

The model selected for this research uses the DME as inputs and the PMESI as the measures of merit, the outputs. Each objective is a function of these PMESI variables and each variable is formed as an index, made of components and subcomponents and aggregated to a $[0, 1]$ scale as described in Saie and Ahner [81]. Every index is composed of open source data derived from the Brookings Institute [71], World Bank [96], CIA factbook [26], and various Department of Defense sources. The remainder of this section describes how this problem is formulated as a dynamic programming problem and solved using rollout algorithms.

4.3.1 Dynamic Programming Requirements.

A general deterministic dynamic programming problem formulation must meet the following requirements:

1. discrete time system of the form $x_{k+1} = f_k(x_k, u_k)$,

2. control constraint, that is, $u_k \in U_k(x_k)$,

3. additive costs of the form $h(x_N) + \sum_{k=0}^{N-1} g(x_k, u_k)$,

4. optimization over (closed-loop) policies, rules for choosing u_k for each k and each possible value of x_k.

The formulation that follows demonstrates each of these requirements.

4.3.2 States.

We define the states as the PMESI variables and describe the state of the system (country) by

$$x_t = \{P_t, M_t, E_t, S_t, I_t\}$$

where the initial state, x_0 is given and the state at any time (in months), $t \in T$ is such that $x_t \in [0, 1]^5$.

4.3.3 Control and Decision Space.

The control space is defined as the DME variables and represent the set of actions an external government can take while conducting nation–building operations. The set $U_t(x_t)$ of feasible controls $u_t(x_t)$ that can be applied to x_t are

$$u_t = \{Dip_t, Mil_t, Eco_t\}$$

where t is in months. The U_t are non–negative real ($\mathbb{R}^+$) numbers and are constrained according to the actual constraints based on the actual data in the problem.

Diplomatic The diplomatic variable (Dip_t) is the percentage of diplomatic assistance where 0 is no assistance (host nation run government), 1 is a government run with external support, and 0.5 represents an approximately equal effort between the two nations. The total amount of diplomatic support must not exceed 70 as the max per time period is 1 and there are 70 time periods, $\sum_{t=0}^{70} Dip_t \leq 70$.

Military The military variable (Mil_t) is the total number of US troops per month (in 100,000s). The maximum number of troops for any t is 1.71 based on the actual data thus $0 \leq Mil_t \leq 1.71 \ \forall \ t$. The total number of troops over the entire time period (70 months) was 98.946 (multiples of 100k) which gives us $0 \leq \sum_{t=0}^{70} Mil_t \leq 98.946$.

Economic The economic variable (Eco_t) is the total amount of aid allocated per month (in billions of US dollars). The maximum aid for any t is \$1.63 billion based on the actual data thus $0 \leq Eco_t \leq 1.63 \ \forall \ t$. The total number of aid over the entire time period (70 months) was 34.15 (in billions of US dollars) yielding $0 \leq \sum_{t=0}^{70} Eco_t \leq 34.15$.

The decision space is continuous, $u_i(t) \in [a, b]^3$, where a and b are constrained based upon the specified constraints for each control. The constraints represent the total amount applied for each resource for a current time period (t) and the entire time period (T). All values are strictly non–negative and are based upon the actual minimum and maximum values that occurred in the first 70 months of operations in Iraq. The diplomatic variable (u_1) is based upon a percentage of diplomatic assistance where 0 is no assistance (host nation run government) and 1 is a government ran by external nation support.

As the decision space contains an infinite number of points for the *Mil* and *Eco* inputs a Nearly Orthogonal and Balanced (NOB) Mixed Design [98] is utilized to effectively explore and evaluate the decision space. This provides 1200 distinct combinations of points to construct, U. A mixed design provides both discrete and continuous factor levels in the design. This chosen design simplifies the decision space, the *Dip* resource is given three possible values, a 0, 0.5, or 1 and the *Mil* and *Eco* resources remain continuous within their constrained values. This prevents the analyst from trying to determine or describe a 37% allocation of the *Dip* resource. The design is built using a spreadsheet [99] which provides the NOB design. At each epoch the resources are applied and the amount available is decremented accordingly. If the resource reaches an amount where the design level for a resource exceeds the amount remaining, that specific combination of controls is not allowable.

4.3.4 System Dynamics.

The system dynamics are represented by the following piecewise differential equation

$$\dot{x}_i(x(t), u(t)) = \begin{cases} 0, & \text{if } x_i(t) + f_i(x(t), u(t)) \leq 0; \\ f_i(x(t), u(t)), & \text{if } 0 < x_i(t) + f_i(x(t), u(t)) < 1; \\ 0, & \text{if } x_i(t) + f_i(x(t), u(t)) \geq 1. \end{cases} \quad (4.1)$$

Where

$$
\begin{aligned}
f_i(x(t), u(t)) &= a_{i1}\left(\frac{P_t}{b_{i1}} - 1\right) + a_{i2}\left(\frac{M_t}{b_{i2}} - 1\right) + a_{i3}\left(\frac{E_t}{b_{i3}} - 1\right) + a_{i4}\left(\frac{S_t}{b_{i4}} - 1\right) \\
&+ a_{i5}\left(\frac{I_t}{b_{i5}} - 1\right) + d_{i1}Dip_t + d_{i2}Mil_t + d_{i3}Eco_t
\end{aligned}
\quad (4.2)
$$

66

for each PMESI index, that is $i = 1, ..., 5$, and the system transitions according to

$$x_{t+1} = x_t + \dot{x}(t) \cdot \delta.$$

The system dynamics (Eq. 4.1) are represented with a piecewise differential equation ensuring the index function remains in its allowable range $[0, 1]$. The values (truncated) for the coefficients described in Equation 4.2 are provided in Tables 4–6.

Table 4. a coefficients

i \ j	Political	Military	Economic	Social	Infrastructure
Political	0.0106437	-0.0221477	0.00031723	-0.0396859	0.00562973
Military	-0.0025072	0.03844424	-0.0036273	-0.0212282	-0.01407430
Economic	-0.0386417	0.03556194	-0.0147303	-0.0773608	-0.01060570
Social	0.00740749	-0.0035113	0.0029288	0.00325991	0.00148458
Infrastructure	-0.0103829	0.01578294	-0.0059614	-0.0141289	-0.00670700

Table 5. b coefficients

i \ j	Political	Military	Economic	Social	Infrastructure
Political	0.32133939	0.94540904	0.05862982	0.97972602	0.32210496
Military	0.04389981	0.99563757	0.20158519	0.18429704	0.43206841
Economic	0.50913874	0.76465212	0.29503016	0.30240614	0.15591104
Social	0.75066697	0.61073867	0.35418124	0.22252805	0.12886480
Infrastructure	0.33070316	0.79409341	0.33970810	0.34378095	0.24653597

4.3.5 Objective Functions.

The nature of the this problem makes it such that a cost is difficult to define and no objective functions exist which are commonly accepted. In this research we present three costs, $c(x_t, u_t)$ to serve as possible objective functions. Since the goal of this problem is improve the state, mathematically we accomplish this by maximizing

Table 6. *d* coefficients

k / i	Diplomatic	Military	Economic
Political	0.00163305	-0.0516518	-0.0005696
Military	-0.0021241	0.15097297	-0.0044780
Economic	-0.0063144	0.17781749	-0.0079398
Social	0.00137478	-0.0097472	0.0003528
Infrastructure	-0.0028405	0.05210194	-0.0013781

the area under the definite integral, or by minimizing the the distance between $x_i(t)$ and 1 with or without assigning a penalty to values further away from 1. All three objective functions are explored.

Reimann Sum $\left(-\sum_{i=1}^{5} f\left(x_i(t)\right) \Delta x_i(t)\right)$, this objective function maximizes the total sum of the indices. Since each index value corresponds with a month (t) and is bounded by $[0, 1]$ they form a definite integral which can be approximated through a Reimann sum. This objective function is referred to as Reimann in Section 4.5.

Penalty Function $\left(\sum_{i=1}^{5}\left(1 - x_i(t)\right)\right)$, this objective function minimizes the sum of the distances between the index value and the max index value of 1. This objective function is referred to as Penalty in Section 4.5.

Squared Penalty Function $\left(\sum_{i=0}^{5}\left(100 - 100 x_i(t)\right)^2\right)$, this objective function is similar to the penalty function. This minimizes the squared distance between index value and the max index value of 1. Each index is scaled by a factor of 100 so the act of squaring the penalty has an increasing affect. This creates a greater penalty for an index value which is further away from 1. This objective function is referred to as Squared Penalty in Section 4.5.

In dynamic programming we consider both the current cost, $c(x_t, u_t)$ and the future costs as given by Bellman's equation

$$J(x_t, u_t) = \min_{u_t} \ c_t(x_t, u_t) + J(x_{t+1}). \qquad (4.3)$$

However, due to the nonlinear system dynamics, we approximate $J(x_{t+1})$ using a heuristic approach to obtain $\widetilde{J}(x_{t+1})$, an approximation.

4.4 Rollout Algorithm

A Rollout Algorithm is an ADP technique to solve problems which fall victim to Bellman's "curse of dimensionality" where there is an exponential increase in computation as the problem size increases. In Section 4.3.2 and 5.3.2 the states space and decision space are defined as Euclidean spaces, $\mathbb{R}^5$ and $\mathbb{R}^3$. In order to achieve computational tractability, the numerical solution of this problem is discretized for each state and control index at each stage. The computation required to carry out each calculation even with this simplification is overwhelming and a closed form solution is not possible. To address this issue a rollout algorithm which makes use of a one-step lookahead scheme and a sub–optimal policy (a base heuristic) which implements the cost–to–go policy of the base heuristic to approximate the optimal cost–to–go is developed. Using this method an approximate control, $\tilde{u}_t$ corresponding to x_t is calculated by $\tilde{J}(x_{t+1})$, an approximation of $J^*(x_{t+1})$. The base heuristic is repeatedly applied at each stage., the system is transitioned to the next time step, and an approximate cost to go $(\tilde{J})$ is calculated. This provides an efficient method to select $\tilde{u}_t(x_t) = \arg\min_{u \in U} c(x_t, u_t) + \widetilde{J}$ where $c(x_t, u_t)$ is the cost–to–go. At every time period, the control which minimizes $(\tilde{u}_t(x_t))$ is selected and applied before the system transitions to the next state.

4.4.1 Base Heuristics.

A total of three base heuristics will be explored each providing an efficient means to calculate $\tilde{J}(x_{t+1})$. In order to increase the importance of improved indices in later periods, weighting is introduced. Each objective function is weighted by the value of t to place more value on future states.

Average – The average policy allocates an equal amount of resources for each future time period. The remaining resources are applied to future states equally. This is accomplished by dividing the remaining resource by the number of remaining time periods $(T - t)$.

Decreasing – This policy allocates resources for each future time period according to a linear decreasing function. A linearly decreasing line from the current time period to the final time period is fit which applies resources in a monotonically decreasing fashion ensuring that all resources are exhausted and the maximum monthly constraint is not violated. This fit dynamically adjusts each time step to calculate $\tilde{J}$ for that decision.

Increasing – This policy allocates resources for each future time period according to a linear increasing function. A linearly increasing line from the current time period to the final time period is fit which applies resources in a monotonically increasing fashion ensuring that all resources are exhausted and the maximum monthly constraint is not violated. This fit dynamically adjusts at each time step to calculate $\tilde{J}$ for that decision.

Algorithm *Rollout Algorithm for the Constrained Nation–Building Problem*

1. Start at $t = 1$.

2. Construct $\tilde{U}$ based on 1200 distinct combinations of points (NOB design) and initial state vector, x_0.

3. **repeat**

4.   For all feasible $\tilde{u}_t \in \tilde{U}$ calculate $x_{t+1} = x_t + \dot{x}(x_t, \tilde{u}_t)$.

5. $\tilde{J}_{t+2} = \sum_{k=t+2}^{T} c(x_k, u_k)$ where the base heuristic is calculated by applying one or a combination of the three heuristics.

6.   Evaluate $c(x_{t+1}, u_{t+1}) + \tilde{J}_{t+2}$.

7. $\tilde{u}_{t+1} = \arg\min_{\tilde{u}} c(x_{t+1}, u_{t+1}) + \tilde{J}_{t+2}$.

8.   Calculate x_{t+1} by applying $\tilde{u}_t(x_t)$.

9.   Increment t by 1.

10. **until** $t = T$

4.5 Results

Using the first 70 months of data from Iraq the algorithm is applied to determine if a better per month allocation strategy exists with the same or less total amount of resources. To determine this, all three heuristics and weighted objective functions were tested, the average runtime (with a 2.5GHz quad-core Intel Core i7) was 171 seconds (standard deviation of 9 seconds) for runs 1-3, 5-7, and 9-11 and 568 seconds (standard deviation of 55 seconds) for runs 4, 8 and 12 as they run each of the heuristics at each time period. The results from each run is provided in Table 7.

From Table 7 we can see that using the rollout ADP approach with both the Reimann sum and the penalty function objective functions demonstrated similar improvement regardless of the heuristic. With runs 1-3 the improvement in the objective function values and the control policies were the same for all 3 runs. The same occurs for runs 9-11 as well. The squared penalty objective function resulted in 3 varying levels of improvement and 3 slightly different control polices, all of the squared penalty policies make no use of the military resource. Generally the PMESI states exhibit a similar pattern- political decreases military, economic, and infrastructure

71

Table 7. Comparison of the Objective Functions and Base Heuristics to the Actual Data

Run	Objective Function	Heuristic	New	Base	% Difference
1	Reimann	Average	193.72	161.26	20%
2	Reimann	Decreasing	193.72	161.26	20%
3	Reimann	Increasing	193.72	161.26	20%
4	Reimann	All	238.31	161.26	48%
5	Squared Penalty	Average	5.90E+07	7.35E+07	20%
6	Squared Penalty	Decreasing	5.47E+07	7.35E+07	26%
7	Squared Penalty	Increasing	5.91E+07	7.35E+07	20%
8	Squared Penalty	All	5.88E+07	7.35E+07	20%
9	Penalty	Average	163.09	186.12	12%
10	Penalty	Decreasing	163.09	186.12	12%
11	Penalty	Increasing	163.09	186.12	12%
12	Penalty	All	148.46	186.12	20%

increase; and social remains in the neighborhood of its start point. In cases where the economic resource is exhausted the military, economic, and infrastructure indices drop drastically and the political increases with no major change to the social index.

When solving the problem using the all 3 heuristics for each objective function the Reimann sum is clearly the best with 48% improvement over all heuristics using that objective function. The Reimann and Penalty objective functions found improved polices based on their objective function values over all of their heuristics applied individually whereas the squared penalty achieves a similar result to the average and decreasing heuristics (run 5 and 6). When comparing the results from the runs using just one heuristic, one can observe that all objective function values increased 12-20% and only the Squared Penalty with the decreasing heuristic provided a different result than the runs with that objective function. Both the Reimann and Penalty objective functions performed identically regardless of the heuristic.

While the optimization functions varied, all presented a significant and consistent improvement. Each of the heuristics and objective functions provided improved results in every case. As an example the state and control plot for run 4 is provided in Figure 8 and 9. From Figure 8 we can observe that the military, economic, and

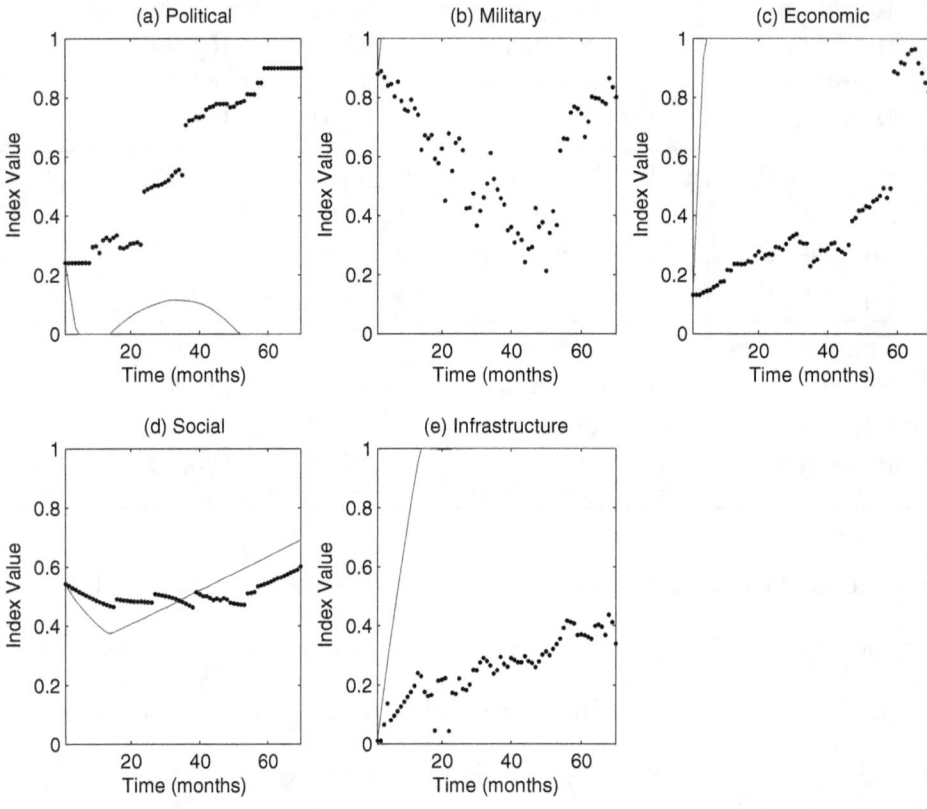

Figure 8. States from run 4 with the Reimann sum objective function– The solid line represents the trajectory provided by the RO solution and the points represent the actual index values for the given time period.

infrastructure indices experienced an immediate and drastic improvement. The social index despite starting lower than the actual values improved over time as well. However, the political index dropped and remained below its start point while achieving an index value of 0 for periods of time during which the other indices continue to improve. The control policy that accompanies these states involved military resources only in the last 20 months with varying levels, economic resources were applied early on and then in the last 20 months, while the diplomatic resources were nearly constantly applied at the highest level, a US lead government. perhaps giving insight to the political state. It is interesting to note that military resources are often needed

73

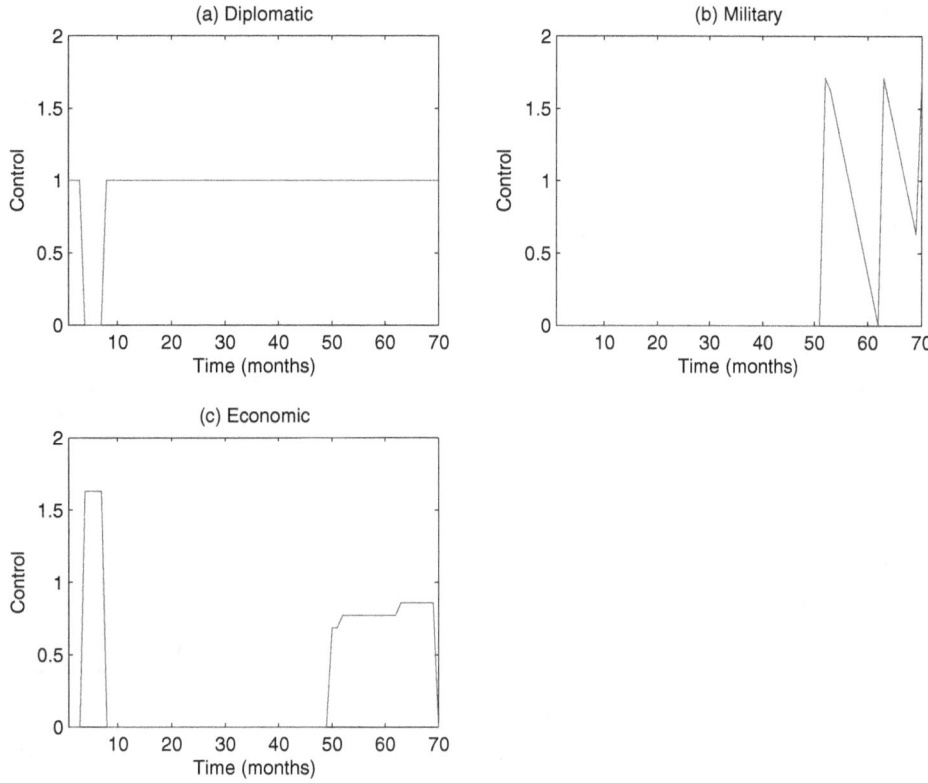

Figure 9. Controls from run 4 with the Reimann sum objective function

to achieve starting conditions for the nation building–problem, in this case it may be impractical to withdraw resources only to reapply later in the time period.

In most of the other cases the level of military resources were low and not frequently applied while higher levels of diplomatic and economic resources were used frequently for prolonged periods of time. The only exception to this is run 12 which made use of all three resources for longer periods of time. The rollout resource allocation used the maximum economic resource 8 times (runs 3-7 and 11-12) and none of the military resources 4 times (runs 2 and 9-11). Overall, the policies generated from the rollout algorithms included significantly fewer military resources, and slightly less of the economic resource. The comparison is found in Table 8.

Overall, this effort demonstrates that the use of rollout algorithms has the potential to improve resource allocation for this nation–building model. Further research

Table 8. Comparison of resource allocation (% less than actual)– The allocation for all 12 runs compared to the actual allocation and the % less resources used.

	1			2			3			4	
Dip	Mil	Eco	Dip	Mil	Eco	Dip	Mil	Eco	Dip	Mil	Eco
66	19.71	22.39	45	0	30.289	44.5	9.09	34.144	34	11.97	34.151
-6%	-80%	-34%	-36%	-100%	-11%	-36%	-91%	0%	-51%	-88%	0%

	5			6			7			8	
Dip	Mil	Eco	Dip	Mil	Eco	Dip	Mil	Eco	Dip	Mil	Eco
34	11.97	34.151	34	11.97	34.151	34	11.97	34.151	34	11.97	34.151
-51%	-88%	0%	-51%	-88%	0%	-51%	-88%	0%	-51%	-88%	0%

	9			10			11			12	
Dip	Mil	Eco	Dip	Mil	Eco	Dip	Mil	Eco	Dip	Mil	Eco
18	0	29.941	23	0	32.522	29	0	34.15	34	11.97	34.151
-74%	-100%	-12%	-67%	-100%	-5%	-59%	-100%	0%	-51%	-88%	0%

may be done with objective functions and the heuristics, as well as implementing additional constraints to prevent rapid military resource changes (i.e. assume groups of troops such as 5,000 per unit and assign them for periods of time).

Implementing a rollout algorithm to determine an improved solution for the allocation of resources was at least 12% and up to 48% better than the actual allocation based on the objective functions considered.

4.6 Conclusion

This research introduces the concept of rollout algorithms to the nation–building problem by demonstrating its applicability through an example using a system of differential equations model and 70 months of data from Iraq. The use of rollout algorithms is shown to present vastly improved policies using various objective functions and base heuristics. Given the complexity of this problem, this approach indicates how these algorithms may be used to address this class of problem by providing policies which improve the state of the country. The results also suggest that this technique may be applied to other types of social–science type problems which require the allocation of limited resources with nonlinear dynamics with similar success.

V. Augmented State Based Violence in Nation–Building Modeling using Approximate Dynamic Programming

5.1 Abstract

Nation–building actions take place amongst the population and as a result, violence is always a concern. The nation–building problem is formulated as a dynamic programming resource allocation problem and modeled with the enemy action of violence. The augmentation of the state vector with the violence factor allows feedback to the state in terms of enemy actions while approximate dynamic programming is applied to allocate how to apply limited resources in a manner that maximizes measurable outcomes or minimize costs. Approximate dynamic programming is implemented through an example consisting of a system of differential equations model of the nation–building operations in Iraq to evaluate the allocation of resources and number of civilian deaths. Multiple cost functions and base heuristics are presented to develop significantly improved policies for given objective functions.

5.2 Introduction

Violence and casualties are often an outcome of armed conflicts, which includes nation–building operations. In the more intense conflicts, such as wars, the number of casualties may be one of several metrics to evaluate success. In earlier times, this often determined the outcome of the battle or war. Lanchester theory [55] is a field of operations research dedicated to this premise and makes use of differential equations to evaluate the strength of two armies as a function of time. In these equations the outcome is determined by the initial strength and effectiveness of each army to determine the rate of change for the enemy (x) and ally (y) strength, by using existing data. Lanchester equations are used to model many conflicts [100, 22, 41, 97, 56] and different types of warfare [55, 78, 32, 84, 85, 37] to measure the population of two forces. These equations are applied to conventional warfare consisting of two armed populations. More recently, conflict consists of battles, not fought in open fields or unpopulated areas but rather in populated areas where civilian casualties are a factor. This is especially true in nation–building operations were the goal is a to improve the state of the country and the majority of actions center around the population. In this case the civilian casualties may be one indicator of success or failure, but regardless if used or not as a metric, casualties influence the state of nation–building. If the number of casualties is low the external and host nation forces can view that as a success, peace is maintained and the people are secure. If that number is high, then it may be viewed as failure with violence prevalent and little or no security for the populace, which may affect the development of a nation. The insurgent or militant force can use civilian casualties to coerce and intimidate people and to disparage the external forces and host nation.

The nation–building problem attempts to model external inputs in order to improve the conditions of the country from a conflict state to a peaceful state through input from an external nation. The nation-building problem (or a sub-set such as counterinsurgency operations) is addressed in the literature through Lanchester equation and differential equation models. In 2008, Blank *et al* [18], developed a dynamic model of insurgency using Lanchester equations and Iraq war data. The model proposes a system of differential equations, the general solution of which is then used to plot the phase portraits of the system and deduce information. Johnson and Madin [47] developed a population model based upon the Logistic differential equation. This model makes use of population size, recruitment, carrying capacity, and mortality to investigate the dynamics in the insurgent population. Schaffer [83] provides a mathematical formulation of 21st Century counterinsurgency warfare using two stochastic time series. Kress and Szechtman [54] model the dynamic relationships among intelligence, collateral casualties in the population, attrition, recruitment to the insurgency, and reinforcement to the government force to show that an insurgency can not be totally eradicate by force, additional actions which affect the attitude of the population are needed as well. Saie and Ahner [81] propose a system of differential equations model to address the nation–building problem using a paradigm based on military planning variables. Here we extend Saie and Ahner [81] to include not only inputs from an external nation but factors (enemy actions) that actively work against moving to a peaceful state.

5.3   Model Including Violence

Since enemy action information is not readily obtainable in nation–building operations, violence in the form of civilian deaths acts a a proxy for these enemy actions within the model. To implement the act of violence in this model Princeton's Empirical Studies of Conflict (ESOC) dataset is used to calculate the number of deaths as a result of insurgent action. From this ESOC dataset we calculate the total number of casualties by month in the first 70 months of the Iraq war. These are only casualties caused by insurgent forces and not coalition forces, which are assumed to be a representation of the current state of the country. The type of casualties concerned are the result of small arms and mortar fire, bombs and other explosive devices, as well as intimidation killings and murder. When some of the casualties for large events are provided using high and low estimates, the sample mean of those estimates is used for modeling. This provides a number of casualties for each month. We build from Section 4.3 and Equation 4.2 by adding another element to the model. Let $V \in \mathbb{Z}^+$ be defined as the number of casualties in a given time period t.

5.3.1 Augmenting the State Vector.

We maintain the states as the PMESI variables and augment the state vector with the new variable, V_t. From Equation 5.3 the key variables in determining the number of casualties are the political and military variables. One can assume that the level of violence in a country will impact the state of that country, so the augmentation of the state vector with the violence is justified.

We now describe the state of the system (country) by

$$x_t = \{P_t, M_t, E_t, S_t, I_t, V_t\}$$

where the initial state, x_0 is given and the state at any time (in months), $t \in T$ is such that $P_t, M_t, E_t, S_t, I_t \in [0,1]^5$ and $V_t \in \mathbb{Z}^+$.

5.3.2 Control and Decision Space.

The control space is defined as the DME variables and represent the set of actions an external government can take while conducting nation–building operations. The set $U_t(x_t)$ of feasible controls $u_t(x_t)$ that can be applied to x_t are

$$u_t = \{Dip_t, Mil_t, Eco_t\}$$

where t is in months. The U_t are non–negative real $(\mathbb{R}^+)$ numbers and are constrained according to the problem. The decision space is continuous, $u_i(t) \in [a,b]^3$, where a and b are constrained based upon the specified constraints for each control. The constraints represent the total amount applied for each resource for a current time period (t) and the entire time period (T). All values are strictly non–negative and are based upon the actual minimum and maximum values that occurred in the first 70 months of operations in Iraq. A Nearly Orthogonal and Balanced (NOB) Mixed Design [98]

80

is implemented to provide 1200 distinct combinations of points to construct, U. This is the same process in described in Chapter IV. At each epoch the resources are applied and the amount available is decremented accordingly. If the resource reaches an amount where the design level for a resource exceeds the amount remaining, that specific combination of controls is not allowable.

5.3.3 System Dynamics.

To include V_t, we rewrite Equation 4.2 as

$$
\begin{aligned}
f_i(x(t), u(t)) &= a_{i1}\left(\frac{P_t}{b_{i1}} - 1\right) + a_{i2}\left(\frac{M_t}{b_{i2}} - 1\right) + a_{i3}\left(\frac{E_t}{b_{i3}} - 1\right) + a_{i4}\left(\frac{S_t}{b_{i4}} - 1\right) \\
&+ a_{i5}\left(\frac{I_t}{b_{i5}} - 1\right) + d_{i1}Dip_t + d_{i2}Mil_t + d_{i3}Eco_t + \gamma_i V_t \quad (5.1)
\end{aligned}
$$

for each PMESI index, that is $i = 1, ..., 5$. The values (truncated) for the coefficients described in Equation 5.1 are provided in Tables 9–11.

Table 9. a coefficients

i \ j	Political	Military	Economic	Social	Infrastructure
Political	0.013479426	-0.02443707	0.000169151	-0.044326194	0.004092453
Military	-0.0175788	0.03585036	-0.00000620	-0.0040145	-0.0062530
Economic	-0.0188127	0.01368835	-0.0107756	-0.0482235	-0.0113442
Social	0.00261637	-0.008244	0.00509257	0.00208774	0.00426975
Infrastructure	-0.0168042	0.01518943	-0.0001110	-0.004930	-0.0025961

5.3.4 Objective Functions.

The nature of the this problem makes it such that a cost is difficult to define and there are no commonly accepted objective functions. In this research we consider the three costs, $c(x_t, u_t)$ described in Chapter IV to serve as objective functions. Since

Table 10. b **coefficients**

j i	Political	Military	Economic	Social	Infrastructure
Political	0.41742117	0.8970741	0.05780014	0.95238897	0.35346225
Military	0.31650255	0.78225326	0.00466155	0.03868765	0.29203887
Economic	0.26296986	0.20580581	0.2920259	0.21512186	0.30590902
Social	0.28936593	0.86394767	0.87821031	0.24140827	0.75522059
Infrastructure	0.56674121	0.54459506	0.00900396	0.17349528	0.17533378

Table 11. d **and** γ **coefficients**

k i	Diplomatic	Military	Economic	Violence
Political	0.00226496	-0.0445818	-0.0006344	-2.13E-06
Military	-0.0033197	0.13759617	-0.0043553	4.04E-06
Economic	-0.0096283	0.14074017	-0.0075998	1.12E-05
Social	0.00200315	-0.0027167	0.00028833	-2.12E-06
Infrastructure	-0.0041684	0.03724438	-0.0012419	4.48E-06

the goal of this problem is improve the state, we accomplish this mathematically by maximizing the area under the definite integral, or by minimizing the the distance between $x_i(t)$ and 1 with or without assigning a penalty to values further away from 1. All three objective functions are explored. In dynamic programming we consider both the current cost, $c(x_t, u_t)$ and the future costs as given by Bellman's equation

$$J(x_t, u_t) = \min_{u_t} \ c_t(x_t, u_t) + J(x_{t+1}). \tag{5.2}$$

However, due to the nonlinear system dynamics, we approximate $J(x_{t+1})$ using a heuristic approach to obtain $\widetilde{J}(x_{t+1})$, an approximation.

5.3.5 Base Heuristics.

A total of three base heuristics (Section 4.4.1) are explored, each providing an efficient means to calculate $\tilde{J}(x_{t+1})$. In order to increase the importance of improved indices in later periods, weighting is introduced. Each objective function is weighted by the value of t to place more value on future states. Each base heuristic is applied to each objective function individually and then all three heuristics are utilized to select $\tilde{u}_t(x_t)$ for each objective function according to the following algorithm. The results are listed in Table 12.

Algorithm *Rollout Algorithm for the Constrained Nation–Building Problem*

1. Start at $t = 1$.

2. Construct $\tilde{U}$ constructed of 1200 distinct combinations of points (NOB design) and initial state vector, x_0.

3. **repeat**

4.   For all feasible $\tilde{u}_t \in \tilde{U}$ calculate $x_{t+1} = x_t + \dot{x}(x_t, \tilde{u}_t)$.

5. $\tilde{J}_{t+2} = \sum_{k=t+2}^{T} c(x_k, u_k)$ where the base heuristic is calculated by applying one or a combination of the three heuristics.

6.   Evaluate $c(x_{t+1}, u_{t+1}) + \tilde{J}_{t+2}$.

7. $\tilde{u}_{t+1} = \arg\min_{\tilde{u}} c(x_{t+1}, u_{t+1}) + \tilde{J}_{t+2}$.

8.   Calculate x_{t+1} by applying $\tilde{u}_t(x_t)$.

9.   Increment t by 1.

10. **until** $t = T$

The results from the model involving the violence factor show that run 4, 6, and 11 are the best for their respective objective functions and run 4 having the most improvement overall. Looking at the resource allocation in Table 13, there were 11 runs which used greater than or equal to 90% fewer military resources and 7

Table 12. Comparison of the Objective Functions and Base Heuristics to the Actual Data Using Actual Violence Data

Run	Objective Function	Heuristic	New	Base	% Difference
1	Reimann	Average	214.62	161.26	33%
2	Reimann	Decreasing	214.62	161.26	33%
3	Reimann	Increasing	214.62	161.26	33%
4	Reimann	All	237.83	161.26	47%
5	Squared Penalty	Average	4.77E+07	7.35E+07	35%
6	Squared Penalty	Decreasing	4.60E+07	7.35E+07	37%
7	Squared Penalty	Increasing	5.32E+07	7.35E+07	28%
8	Squared Penalty	All	5.31E+07	7.35E+07	28%
9	Penalty	Average	162.10	186.12	13%
10	Penalty	Decreasing	162.10	186.12	13%
11	Penalty	Increasing	160.14	186.12	14%
12	Penalty	All	166.14	186.12	11%

Table 13. Comparison of resource allocation (% less than actual)– The allocation for all 12 runs compared to the actual allocation and the % less resources used with actual violence data.

	1			2			3			4	
Dip	Mil	Eco	Dip	Mil	Eco	Dip	Mil	Eco	Dip	Mil	Eco
61.5	1.71	16.301	51	0	30.632	41	9.63	34.144	31	1.71	34.139
-12%	-98%	-52%	-27%	-100%	-10%	-41%	-90%	0%	-56%	-98%	0%

	5			6			7			8	
Dip	Mil	Eco	Dip	Mil	Eco	Dip	Mil	Eco	Dip	Mil	Eco
32	3.42	34.139	32	3.42	34.139	32	3.42	34.139	31	1.71	34.139
-54%	-97%	0%	-54%	-97%	0%	-54%	-97%	0%	-56%	-98%	0%

	9			10			11			12	
Dip	Mil	Eco	Dip	Mil	Eco	Dip	Mil	Eco	Dip	Mil	Eco
48.5	0	39.003	48.5	0	32.946	51	0	26.658	32	1.62	34.138
-29%	-100%	-15%	-31%	-100%	-4%	-27%	-100%	-16%	-54%	-98%	0%

runs which used the same allocation of economic resources. The run with the most improved objective function (run 4) observed both a significant decrease in military resources and used all of the economic resources. Each of the heuristics and objective functions provided improved results in every case. As an example the state and control plot for run 4 is provided in Figure 10 and 11. Overall adding the violence

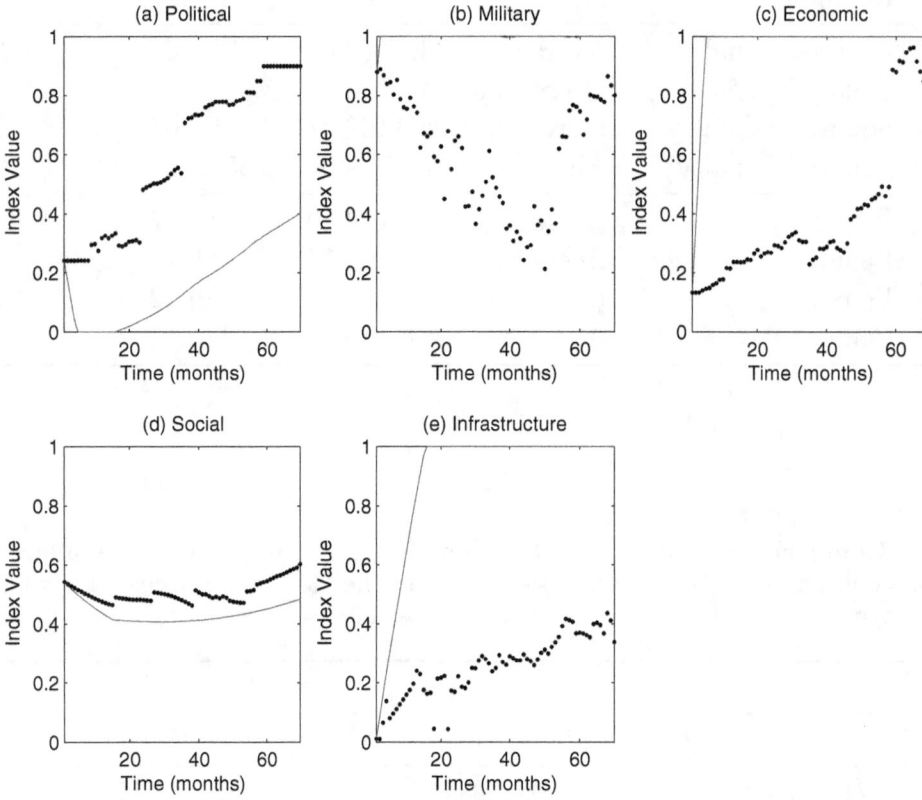

Figure 10. States from run 4 with the Reimann sum objective function— The solid line represents the trajectory provided by the RO solution and the points represent the actual index values for the given time period.

factor to the model decreased the best and worst percent difference for a run by 1% but increased the average and decreasing heuristic runs by 8-15%. The addition of this factor does not degrade or confound the model in any way.

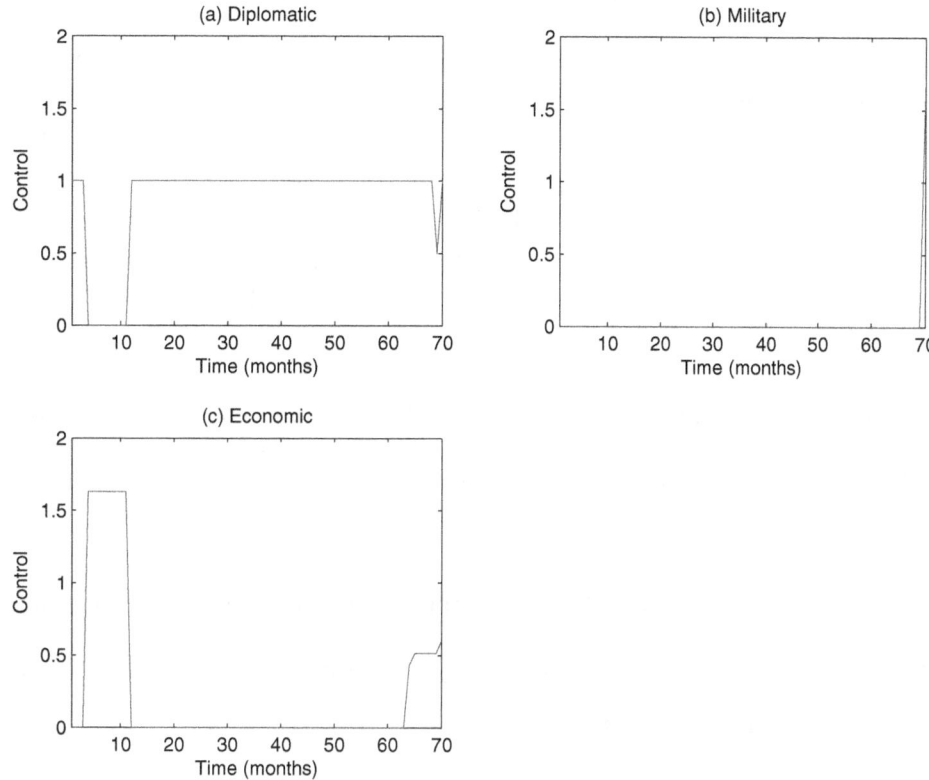

Figure 11. Controls from run 4 with the Reimann sum objective function

5.4   Model with Predicted Violence

To account for the element of violence in this model we consider what PMESI factors contribute to violence. To accomplish this we conduct a stepwise regression using the PMESI index values as the predictor variables and the casualty data as the response. A factorial to degree two design is selected to test all main and two-way interaction effects. The resulting model

$$\widetilde{V}_t = 1729.284 + 1331.622 * P_t - 2458.36 * M_t - 6088.69\left(\left(P_t - 0.58991\right)\left(M_t - 0.60008\right)\right)$$

(5.3)

is determined using the minimum Bayesian Information Criterion by forward regressing on the predictor variables. All variables are significant as shown by the *p–values* in Table 14. Of particular note, the political and military variables are the only PMESI

86

variables entering the model and resulted in an adjusted R^2 of 0.892. The dataset has

Table 14. Parameter Estimates

Term	Estimate	Std Error	t Ratio	p–value
Intercept	1729.284	162.9916	10.61	< 0.0001
P_t	1331.622	145.421	9.16	< 0.0001
M_t	-2458.36	162.8202	-15.1	< 0.0001
$(P_t - 0.58991) * (M_t - 0.60008)$	-6088.69	767.3112	-7.94	< 0.0001

range of 2,380, mean of 1114.23, and a standard deviation of 701.88 indicating a high level of variability in the data. Initial regression failed verification of the normality of errors assumption at the tails of the data. To reduce the variance and influence of outliers the violence data is transformed using the natural logarithm function; additionally the first month is excluded as it is not a complete month (March 30-31 2003). The Chi–Squared and Shapiro–Wilk goodness of fit tests are conducted to determine if the residuals are normally distributed. Both tests indicate normality with *p-values* of 0.0956 and 0.1774 respectively with an $\alpha = 0.05$. Based on the *p-values* in Table 14 and from the goodness of fit tests one can conclude that the predictor model for V is appropriate.

As the rollout algorithms find improved objective function values the corresponding state vector also changes. As $\widetilde{V}$ is derived from the current state, the level of violence will also change as the state changes. Incorporating $\widetilde{V}$ into Equation 5.1 now will determine the level of violence as a function of the current state vector and the calculated level of violence is used instead of the actual. The future states will now include $\widetilde{V}$ in determining the rate of change. Once again this is modeled in the same manner as Chapter IV using the rollout algorithms and the objective function values are compared to the values from the actual data. The results are listed in Table 15.

Table 15. Comparison of the Objective Functions and Base Heuristics to the Actual Data with Calculated Violence Data

Run	Objective Function	Heuristic	New	Base	% Difference
1	Reimann	Average	209.89	161.26	30%
2	Reimann	Decreasing	209.89	161.26	30%
3	Reimann	Increasing	209.89	161.26	30%
4	Reimann	All	239.21	161.26	48%
5	Squared Penalty	Average	5.03E+07	7.35E+07	32%
6	Squared Penalty	Decreasing	4.84E+07	7.35E+07	34%
7	Squared Penalty	Increasing	5.57E+07	7.35E+07	24%
8	Squared Penalty	All	5.55E+07	7.35E+07	24%
9	Penalty	Average	164.55	186.12	12%
10	Penalty	Decreasing	165.55	186.12	12%
11	Penalty	Increasing	163.39	186.12	12%
12	Penalty	All	168.90	186.12	9%

Just as in previous runs, the improvement of runs 4 and 6 improved the most for their respective objective function, while run 10 is the best for the Penalty objective function, and run 4 is the most improved overall. Looking at the resource allocation in Table 13 there were 11 runs which used greater than or equal to 97% fewer military resources and 7 runs which used the same allocation of economic resources. The run with the most improved objective function (run 4) observed both significant decrease in military resources and the use of all economic resources. All presented a significant and consistent improvement. Each of the heuristics and objective functions provided improved results in every case. As an example the state and control plot for run 4 is provided in Figure 12 and 13. Overall adding the violence factor to the model decreased worst run percent difference by 3% and the best run percent remained at 48%. The remainder remained at the same percent difference or increased up to 12%. The addition of this factor does not degrade or confound the model in any way.

Table 16. Comparison of resource allocation (% less than actual)– The allocation for all 12 runs compared to the actual allocation and the % less resources used with calculated data.

	1			2			3			4	
Dip	Mil	Eco	Dip	Mil	Eco	Dip	Mil	Eco	Dip	Mil	Eco
61.5	1.71	16.645	51.5	0	30.543	34	16.38	34.142	32	1.17	34.141
-12%	-98%	-51%	-26%	-100%	-11%	-51%	-83%	0%	-54%	-99%	0%

	5			6			7			8	
Dip	Mil	Eco	Dip	Mil	Eco	Dip	Mil	Eco	Dip	Mil	Eco
33	2.88	34.141	33	2.88	34.141	33	2.88	34.141	32	1.17	34.141
-53%	-97%	0%	-53%	-97%	0%	-53%	-97%	0%	-54%	-99%	0%

	9			10			11			12	
Dip	Mil	Eco	Dip	Mil	Eco	Dip	Mil	Eco	Dip	Mil	Eco
48	0	30.123	33.5	0	32.693	51.5	0	29.863	31	1.71	34.141
-31%	-100%	-12%	-52%	-100%	-4%	-26%	-100%	-13%	-56%	-98%	0%

5.4.1 Violence as a augmented state variable.

Violence is considered a random variable so that the actual data is a probabilistic outcome. In this section, the calculated violence is used in place of the actual data. In Figure 14 we can first observe that the predicted violence level, $\widetilde{V}$ based on the actual states is an adequate fit to the actual violence data. Additionally two very distinct trajectories of violence are observed, one which deaths increases drastically and one which deaths decrease to lower levels. In runs 3, 11, and 12 the number of casualties rapidly increase from 500 to 4500 at months 50-60 corresponding to the military state dropping and the depletion of economic resources. In runs 1, 2, 4-6, 9, and 10 the level of violence remains generally constant with a slight decline, where as in runs 7 and 8 we see a decline to a level very close or at 0. In both run 7 and 8 the military state remained at a high level (near 1) and the economic resources were allocated throughout the entire time period.

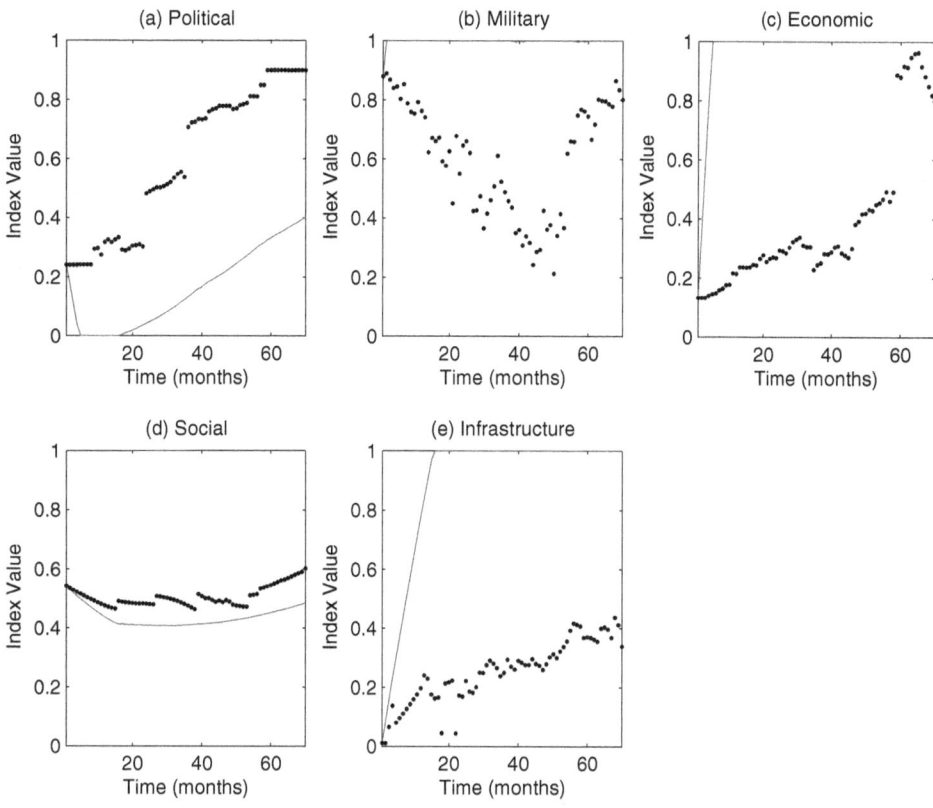

Figure 12. States from run 4 with the Reimann sum objective function– The solid line represents the trajectory provided by the RO solution and the points represent the actual index values for the given time period.

5.4.2 Violence as a Function of PMESI.

The implementation of Equation 5.3 demonstrates two import aspects. First, $\widetilde{V}_t$ is described as a function of PMESI, specifically the political and military variables. This allows the model to continually update the number of civilian deaths based upon the state vector providing instantaneous feedback to the model at each epoch. Secondly, this allows for limited predictive capabilities based upon the state of country whereas most models deal with military casualties where military capability, tactics, and protection play an integral role in determining the number of deaths. As a area for future work a single objective function to minimize deaths may be implemented.

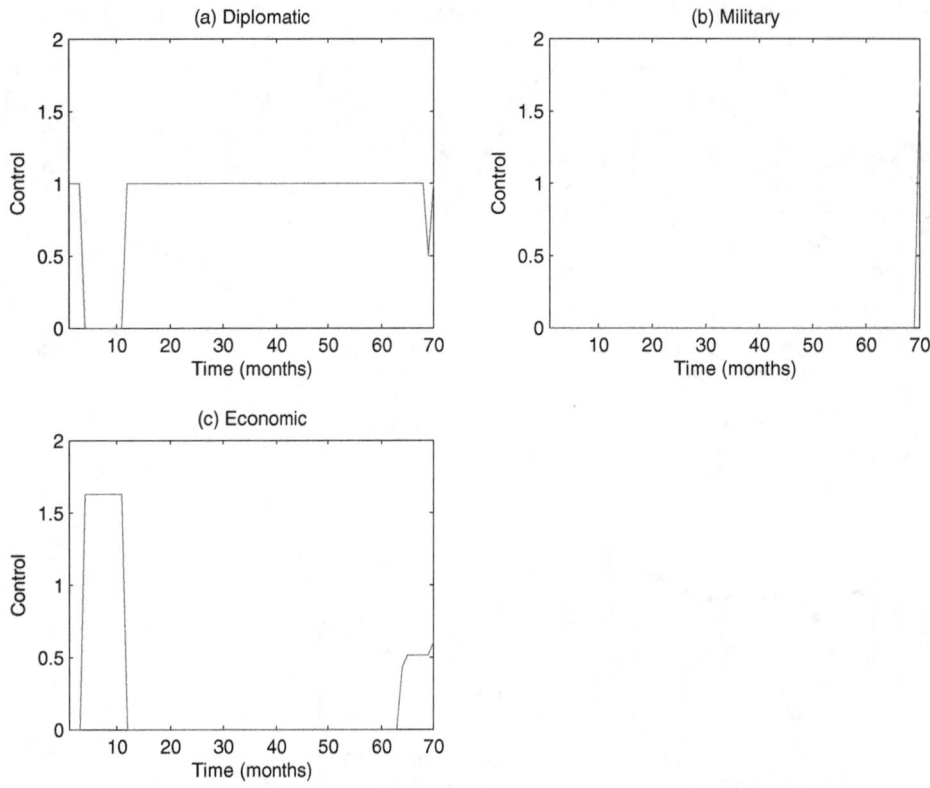

Figure 13. Controls from run 4 with the Reimann sum objective function

5.5   Conclusions

The addition of feedback to the model via violence highlights a few key points. First, augmenting the state with the violence factor overall improves the dynamic programming model and approach. This allows the rollout algorithms in general to find solutions which improve the objective function. While the maximum and minimum improvement in Table 12 did decrease by 1% the majority of the improvement was 8-13% higher than in Table 7. Similar results are found when implementing the calculated violence. The inclusion of the calculated violence versus the actual violence provides a means to base the violence on the PMESI state of the country. Secondly, it demonstrates the use of determining violence through civilian casualties and the state of the country rather than levels and capabilities of armed forces. This is important

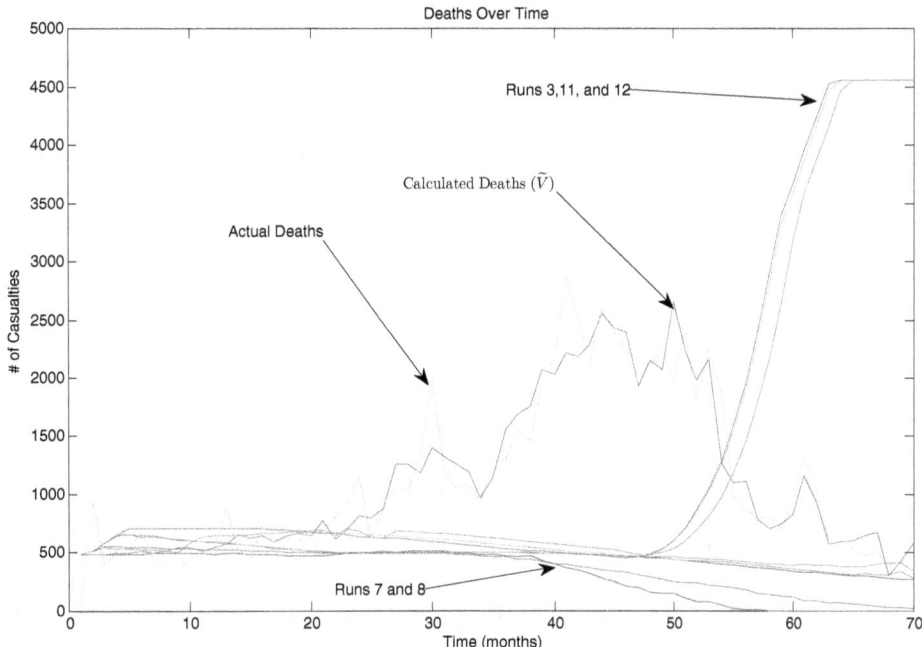

Figure 14. Casualties over time

in nation–building operations and other non–kinetic operations where the primary mission of the armed force is not offensive in nature.

VI. Contributions and Conclusion

Throughout this research several conclusions and contributions were demonstrated. This chapter presents the major findings.

6.1 Methodology and Model

A modeling methodology is developed which creates indices to capture the "state" of a nation. This method is novel and innovative in that it makes use of open source data and is adaptable to the set of available data and is accomplished using the DIME–PMESII paradigm. The data aspect of this model makes use of various open source data, which is available for several nations. The DIME-PMESII paradigm provides a framework to describe the states and controls of this complex system. This framework is easily adaptable to any nation and will make use of the available data. The ability of the model is illustrated through the nation–building effort in Iraq. Additionally, support and interest in the model has been expressed by several commands (Central Command, US Army Africa, and Pacific Command) for wargaming and planning purposes. Generation of this model provides a foundation on which to expand and the dynamics required to generate resource allocation policies.

6.2 Use of Dynamic Programming

Dynamic programming provides a framework in which to use the developed model to solve this political and social science problem and elicit control policies which improve the state of the system. Several different objectives and heuristics were tested using rollout algorithms with improvement over the actual policy in every case. While this is the initial work to frame the problem using a dynamic programming framework, additional work may be done with the objective functions, heuristics, and constraints to enhance the model.

6.3 Inclusion of Violence

One of the most unfortunate aspects of all phases of war is the death of civilians; this aspect is modeled in two ways. First, through inclusion within the model of actual data collected to measure these incidents and then through an expected measure of violence. Implementing the violence factor provides feedback to the model in the form of enemy or insurgent force's actions. The expected measure is novel in that it makes use of the PMESI state of the country to determine the level of violence. The future state is determined through the rate of change based on the current state, controls, and violence. Additional work may be done to implement the number of deaths as an objective or part of a multi–objective approach.

6.4 Conclusion

Overall a novel, traceable, and defendable approach to the nation–building problem is developed and implemented to address a gap in both modeling and resource allocation to political and social science problems using dynamic programming. The testing of multiple objectives and use of heuristics through rollout algorithms show policies exist which improve the state according to the objective function value. The implementation of enemy action in the model enhances the model and more accurately defines the system according to the real world problem it is modeling. Improvement is shown by generating policies to increase an objective function based on a nation's PMESI state variables. This modeling approach and implementation of dynamic programming provides a significant ability to wargamers and modelers concerned with the nation–building problem.

6.5 Future Work

This initial modeling effort, use of a dynamic programming framework, and development of rollout algorithms for resource allocation offer a rich foundation to for future research to include:

- PMESI index generation strategies as new data are available.

- Additional methods to determine the coefficients in the system of differential equations.

- Stability analysis of the system of differential equations.

- Poisson distributed casualties as a function of the PMESI state.

- Augmenting the state with other enemy actions and SME input.

- Additional objective functions which may include violence or other enemy actions.

- Heuristics to better approximate cost–to–go function.

- Explore other conjectured functional forms for the system of differential equations.

- More refined numerical differential equation solution methods.

Appendix A. Weights and PMESI-DME values

Table 17. Weights and PMESI values

Weights	P	M	E	S	I
10	0.240586	0.879987	0.132659	0.542361	0.01059
10	0.240653	0.889444	0.132659	0.535181	0.009889
10	0.240718	0.868313	0.132659	0.528231	0.06475
10	0.240781	0.840435	0.139791	0.521506	0.137319
10	0.240844	0.845627	0.145498	0.515003	0.07991
10	0.240905	0.803333	0.148351	0.508717	0.09553
10	0.240966	0.853037	0.158693	0.502644	0.111153
10	0.241025	0.788425	0.164399	0.496779	0.126978
10	0.294386	0.759259	0.175811	0.49112	0.143145
10	0.297033	0.753802	0.177595	0.485662	0.159568
1	0.274542	0.792581	0.216499	0.480401	0.176159
1	0.317623	0.763066	0.214697	0.475332	0.196826
1	0.325756	0.741637	0.236789	0.470867	0.240652
1	0.316681	0.623019	0.236771	0.467062	0.229184
1	0.325994	0.671914	0.235682	0.463851	0.176061
1	0.333737	0.660867	0.236733	0.490333	0.162521
1	0.292175	0.672715	0.244917	0.488105	0.165584
1	0.289083	0.592473	0.243115	0.486268	0.044565
1	0.294907	0.5772	0.265208	0.484754	0.213742
1	0.304837	0.626966	0.277671	0.483497	0.216987
1	0.306734	0.449628	0.255184	0.48243	0.222815
1	0.310245	0.678609	0.265865	0.481487	0.043264
1	0.301896	0.550076	0.270487	0.482044	0.173015
1	0.481669	0.646522	0.267977	0.481107	0.169039
1	0.489428	0.66091	0.293998	0.480093	0.221532
1	0.495546	0.621693	0.291131	0.478935	0.185926
1	0.502322	0.423912	0.284698	0.5069	0.18195
1	0.502403	0.426045	0.303943	0.505254	0.2011
1	0.506309	0.473811	0.322474	0.503264	0.250048
1	0.513206	0.365226	0.33102	0.500863	0.2486
1	0.520224	0.415487	0.337782	0.497987	0.275364
1	0.534915	0.460654	0.310664	0.494569	0.290815
1	0.547781	0.507261	0.305301	0.490545	0.280312
1	0.555534	0.612021	0.305644	0.485876	0.265316
1	0.537418	0.522916	0.228721	0.480629	0.238117
1	0.707723	0.487344	0.244274	0.474901	0.249329
1	0.722704	0.4576	0.251279	0.468783	0.294008

Continued on Next Page...

Table 17 – Continued

Weights	P	M	E	S	I
1	0.724949	0.436149	0.28222	0.462372	0.27029
1	0.734419	0.349506	0.281843	0.513929	0.260553
1	0.733153	0.360201	0.288689	0.507213	0.29037
1	0.736864	0.308068	0.304922	0.500486	0.283341
1	0.759876	0.338772	0.308448	0.500509	0.275967
1	0.767478	0.31686	0.285379	0.494041	0.275778
1	0.770784	0.242006	0.277458	0.48784	0.296197
1	0.779	0.286448	0.270078	0.491999	0.280122
1	0.778518	0.293023	0.300522	0.486608	0.274002
1	0.779445	0.424931	0.38192	0.493427	0.259616
1	0.779117	0.361439	0.391993	0.489215	0.278725
1	0.767905	0.376936	0.417044	0.478021	0.301756
1	0.769977	0.212308	0.41927	0.475483	0.312836
1	0.781398	0.340717	0.432195	0.473563	0.299445
1	0.784269	0.413797	0.428002	0.472246	0.321493
1	0.788793	0.367303	0.448774	0.471515	0.337761
1	0.812236	0.620004	0.454566	0.509688	0.355672
1	0.811699	0.660914	0.467492	0.511723	0.392405
1	0.811161	0.659484	0.492899	0.514252	0.416556
1	0.849936	0.748501	0.460532	0.533882	0.412167
1	0.850014	0.768705	0.491289	0.537216	0.407258
1	0.899504	0.762446	0.888588	0.540861	0.36836
1	0.899507	0.745153	0.880733	0.544754	0.37034
10	0.89951	0.666884	0.918171	0.550241	0.367233
10	0.899513	0.718053	0.913883	0.555852	0.361684
10	0.899516	0.802456	0.947041	0.561528	0.355089
10	0.899519	0.798139	0.960942	0.563872	0.399391
10	0.899522	0.796374	0.96343	0.56949	0.403605
10	0.899526	0.786499	0.916702	0.574987	0.396189
10	0.899529	0.778416	0.882457	0.580301	0.368093
10	0.899532	0.865277	0.849995	0.585371	0.436651
10	0.899535	0.834748	0.819316	0.590135	0.411707
10	0.899538	0.801217	0.793273	0.602032	0.338448

Table 18. DME values

Dip	Mil	Eco
0.0000	0	0.428571
1.0000	1.5	0.428571
1.0000	1.5	0.428571
1.0000	1.49	0.428571
1.0000	1.39	0.428571
1.0000	1.32	0.428571
1.0000	1.31	0.428571
1.0000	1.23	1.625
1.0000	1.22	1.625
1.0000	1.22	1.625
1.0000	1.15	1.625
1.0000	1.3	1.625
1.0000	1.37	1.625
1.0000	1.38	1.625
1.0000	1.38	1.625
0.5000	1.4	1.625
0.5000	1.4	1.625
0.5000	1.38	1.625
0.5000	1.38	1.625
0.5000	1.38	0.166667
0.5000	1.48	0.166667
0.5000	1.5	0.166667
0.5000	1.55	0.166667
0.5000	1.5	0.166667
0.5000	1.42	0.166667
0.5000	1.38	0.166667
0.5000	1.35	0.166667
0.5000	1.38	0.166667
0.5000	1.38	0.166667
0.5000	1.38	0.166667
0.5000	1.52	0.166667
0.5000	1.6	0.266667
0.5000	1.6	0.266667
0.5000	1.36	0.266667
0.5000	1.33	0.266667
0.5000	1.33	0.266667
0.5000	1.32	0.266667
0.5000	1.32	0.266667
0.5000	1.269	0.266667
0.5000	1.3	0.266667

Continued on Next Page...

Table 18 – Continued

Dip	Mil	Eco
0.5000	1.38	0.266667
0.5000	1.44	0.266667
0.5000	1.44	0.266667
0.5000	1.4	0.266667
0.5000	1.4	0.266667
0.5000	1.32	0.266667
0.5000	1.35	0.266667
0.5000	1.42	0.266667
0.5000	1.46	0.266667
0.5000	1.497	0.266667
0.5000	1.57	0.266667
0.5000	1.6	0.266667
0.5000	1.62	0.266667
0.5000	1.68	0.266667
0.5000	1.71	0.266667
0.5000	1.62	0.225
0.5000	1.6	0.225
0.5000	1.57	0.225
0.5000	1.57	0.225
0.5000	1.55	0.225
0.5000	1.53	0.225
0.5000	1.5	0.225
0.5000	1.48	0.225
0.5000	1.48	0.225
0.5000	1.48	0.225
0.5000	1.48	0.225
0.5000	1.48	0.225
0.5000	1.48	0.183333
0.5000	1.45	0.183333
0.5000	1.42	0.183333

Appendix B. Civilian Deaths by Month

Table 19. Civilian Deaths by Month

Month	Deaths	Month	Deaths	Month	Deaths
1	36	31	1190	61	1334
2	931	32	1052	62	1025
3	382	33	1090	63	608
4	523	34	917	64	600
5	542	35	1280	65	525
6	678	36	1286	66	474
7	448	37	1561	67	505
8	390	38	1462	68	392
9	361	39	1927	69	420
10	453	40	2177	70	391
11	484	41	2866		
12	544	42	2519		
13	874	43	2111		
14	523	44	2609		
15	447	45	2387		
16	720	46	2225		
17	625	47	2173		
18	519	48	2101		
19	590	49	2274		
20	698	50	1981		
21	673	51	2392		
22	808	52	1811		
23	901	53	2239		
24	1151	54	1850		
25	619	55	992		
26	874	56	852		
27	1045	57	790		
28	1031	58	795		
29	1337	59	711		
30	1988	60	907		

Appendix C. State and Control Plots for Chapter IV

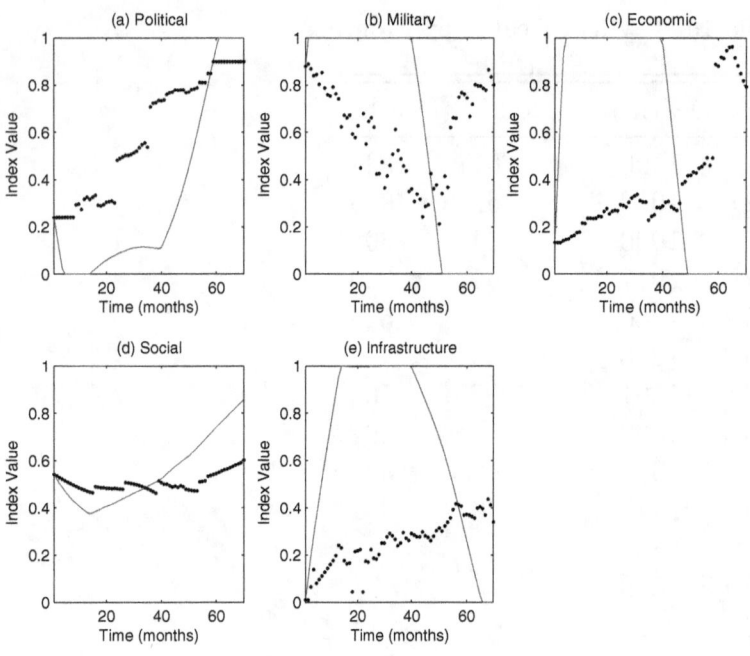

Figure 15. Run 1 States for Chapter IV

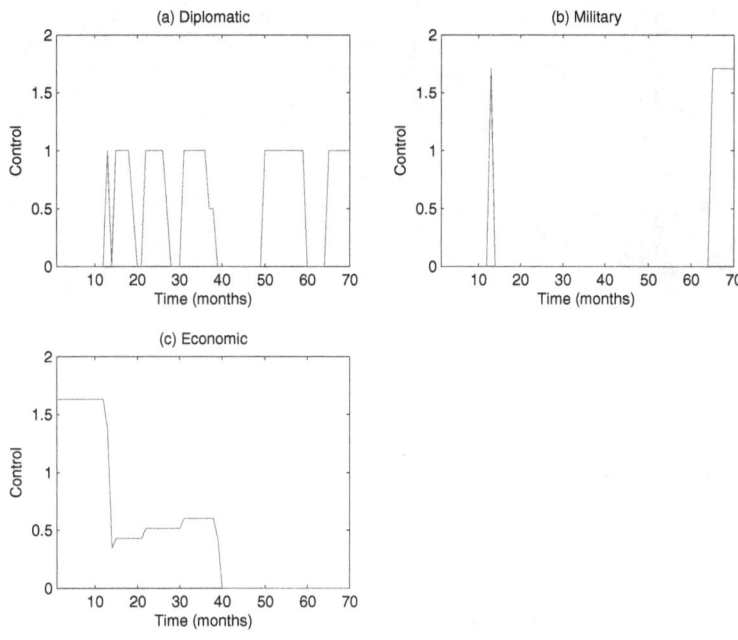

Figure 16. Run 1 Controls for Chapter IV

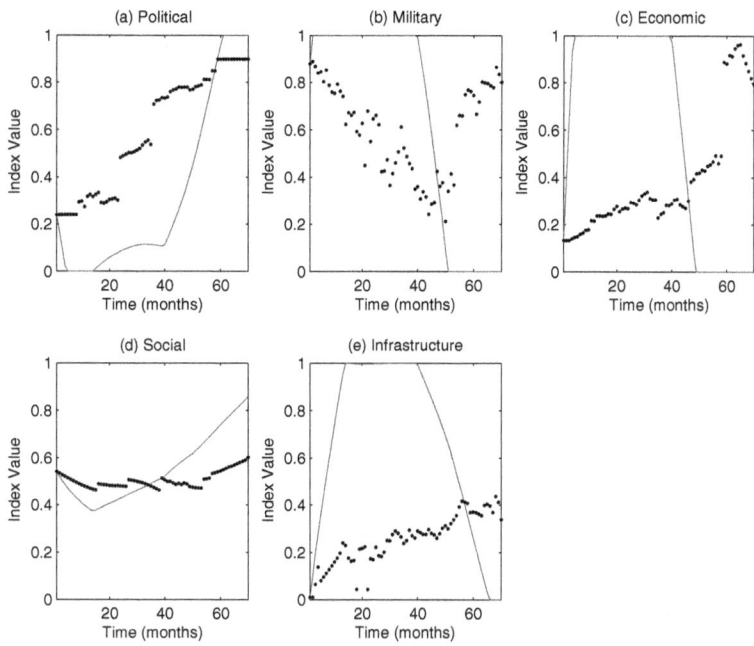

Figure 17. Run 2 States for Chapter IV

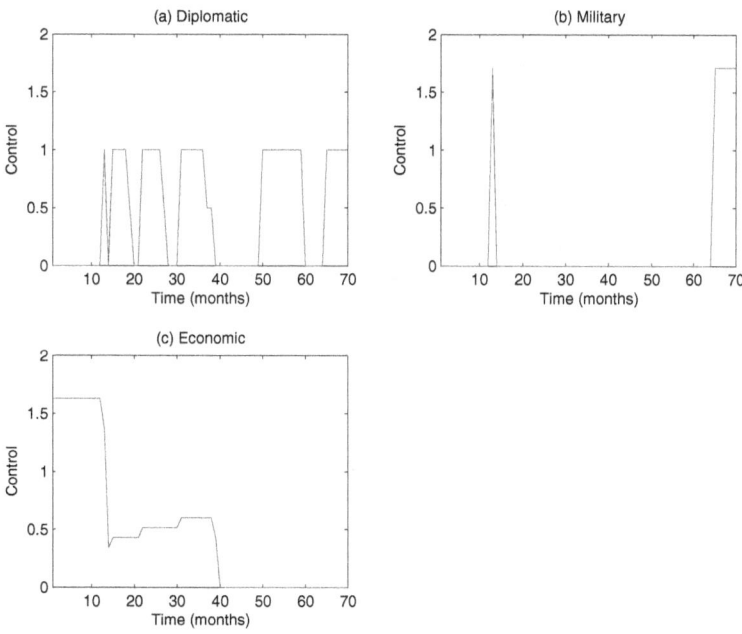

Figure 18. Run 2 Controls for Chapter IV

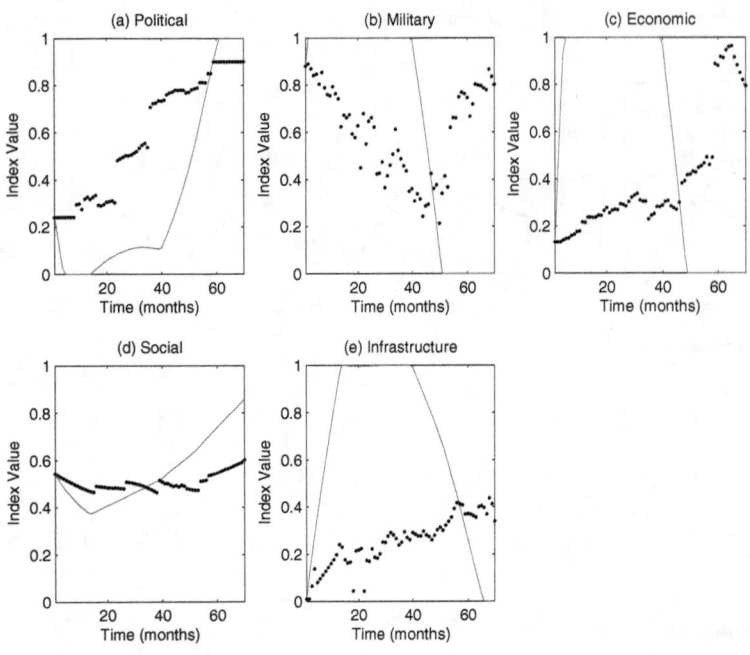

Figure 19. Run 3 States for Chapter IV

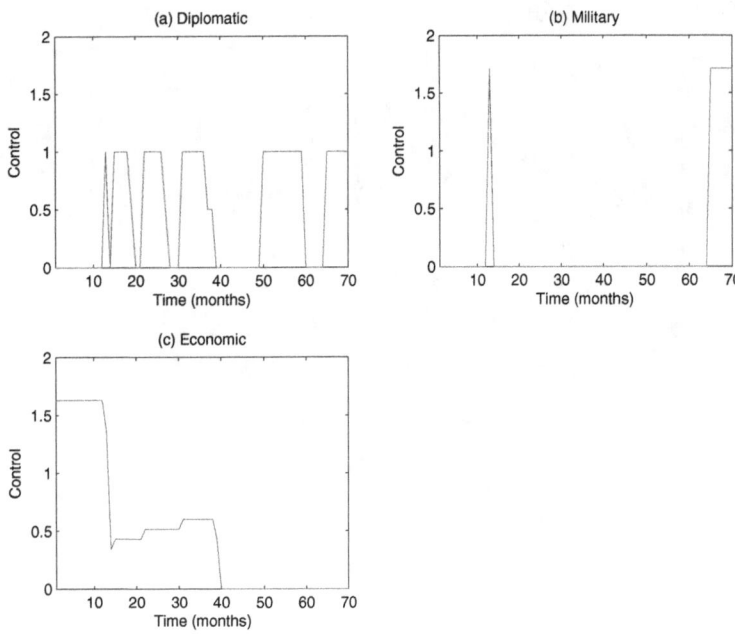

Figure 20. Run 3 Controls for Chapter IV

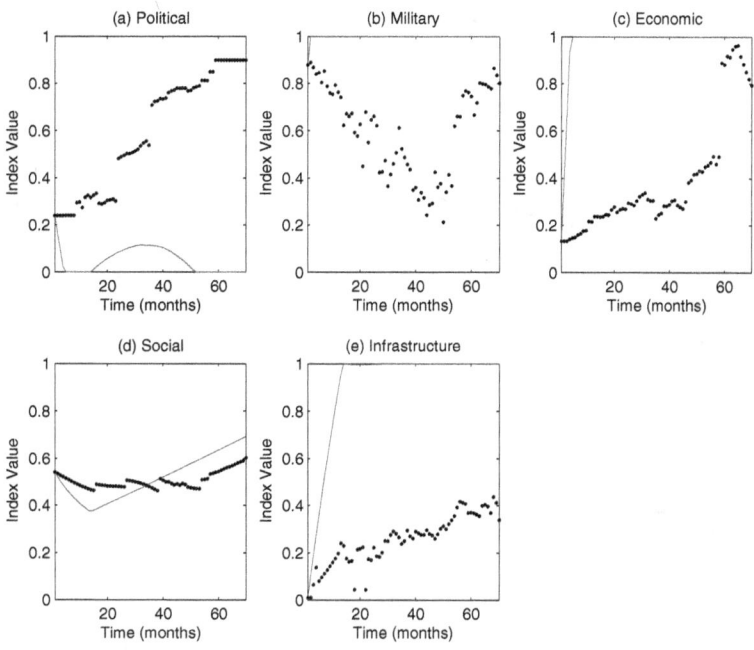

Figure 21. Run 4 States for Chapter IV

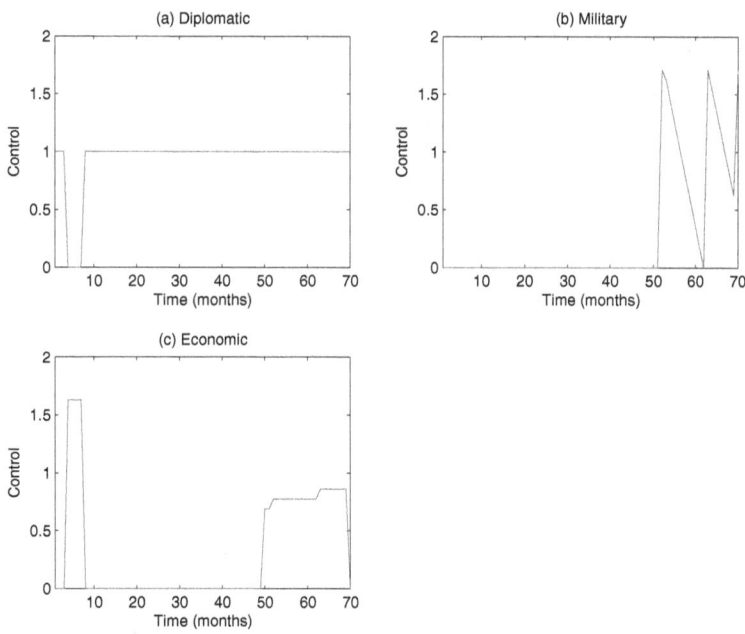

Figure 22. Run 4 Controls for Chapter IV

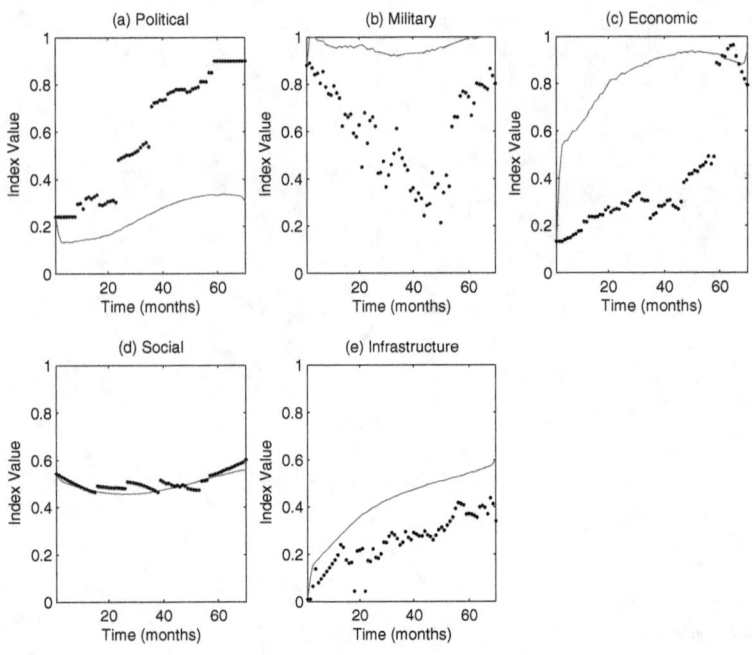

Figure 23. Run 5 States for Chapter IV

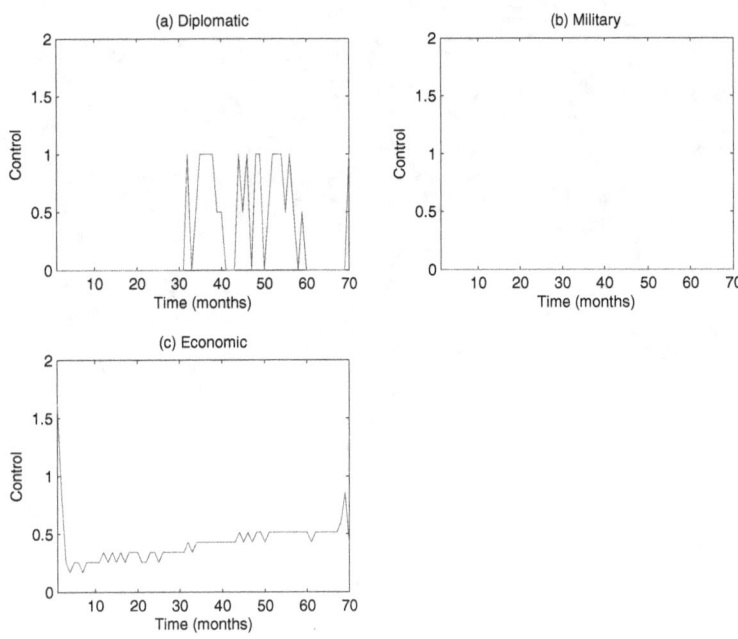

Figure 24. Run 5 Controls for Chapter IV

105

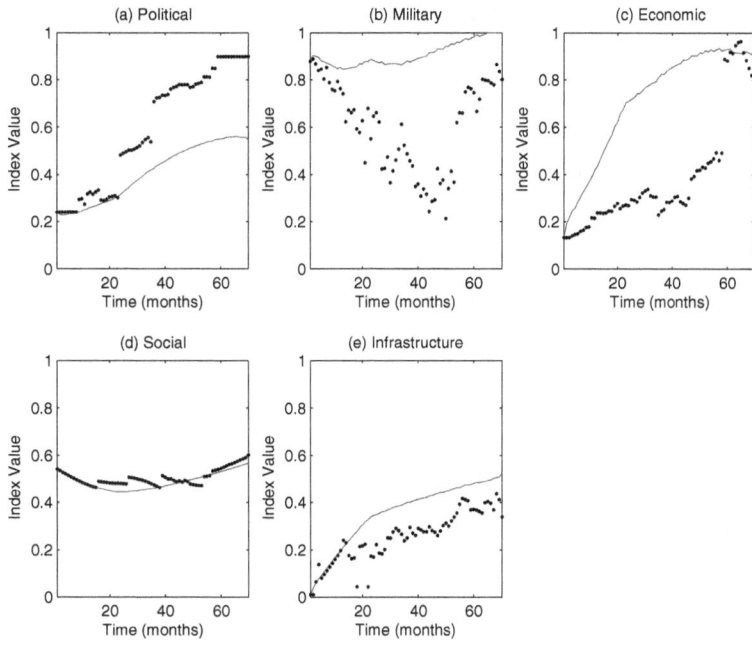

Figure 25. Run 6 States for Chapter IV

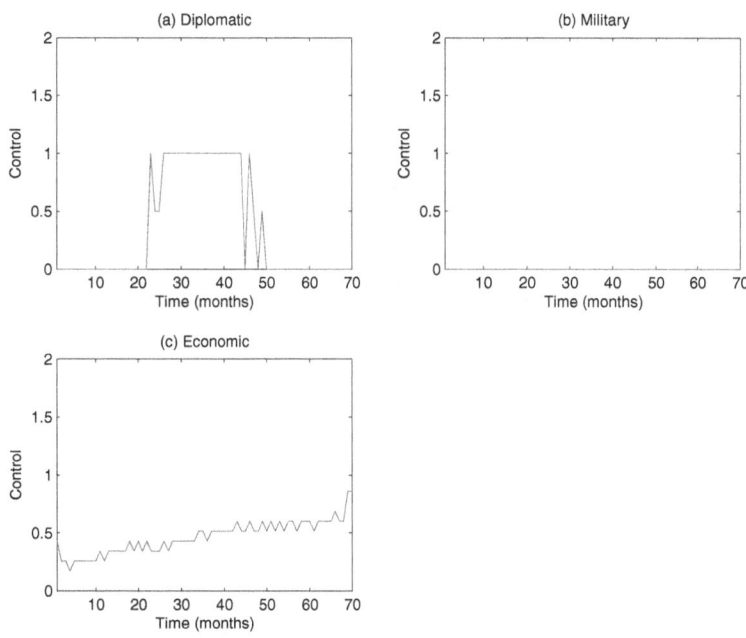

Figure 26. Run 6 Controls for Chapter IV

106

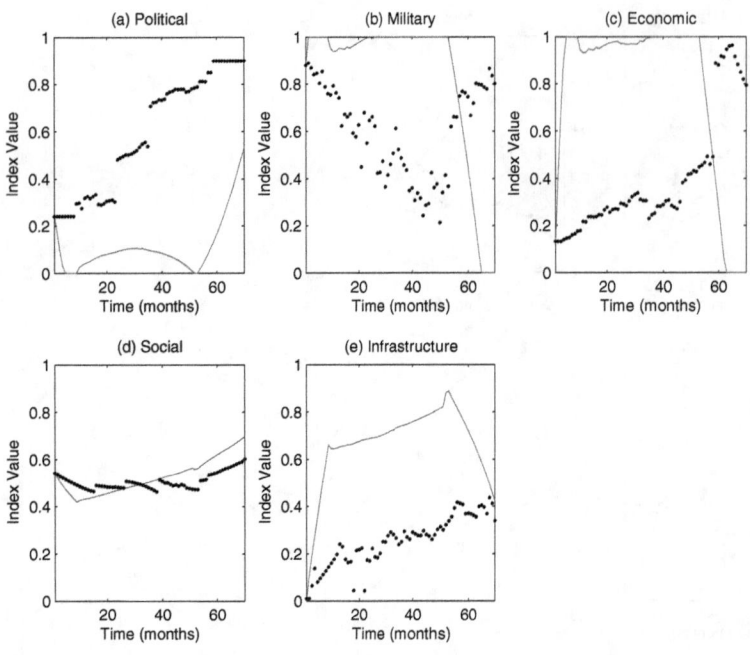

Figure 27. Run 7 States for Chapter IV

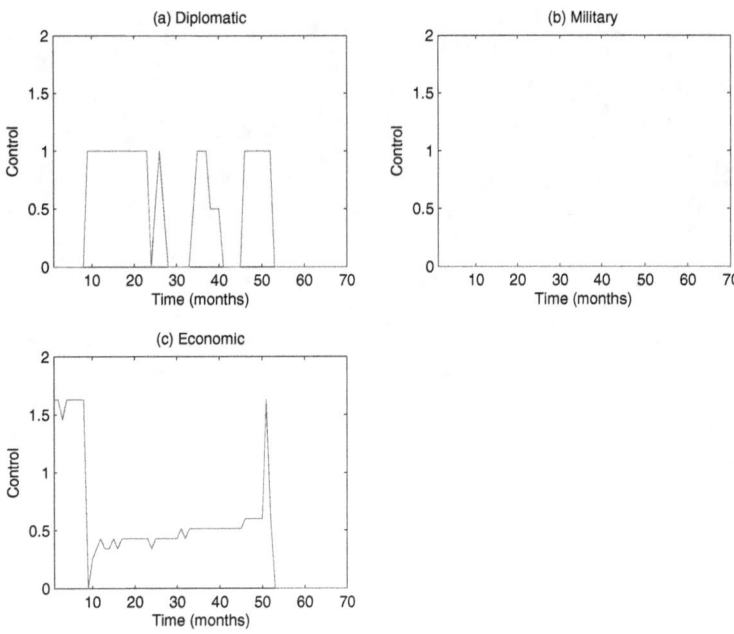

Figure 28. Run 7 Controls for Chapter IV

107

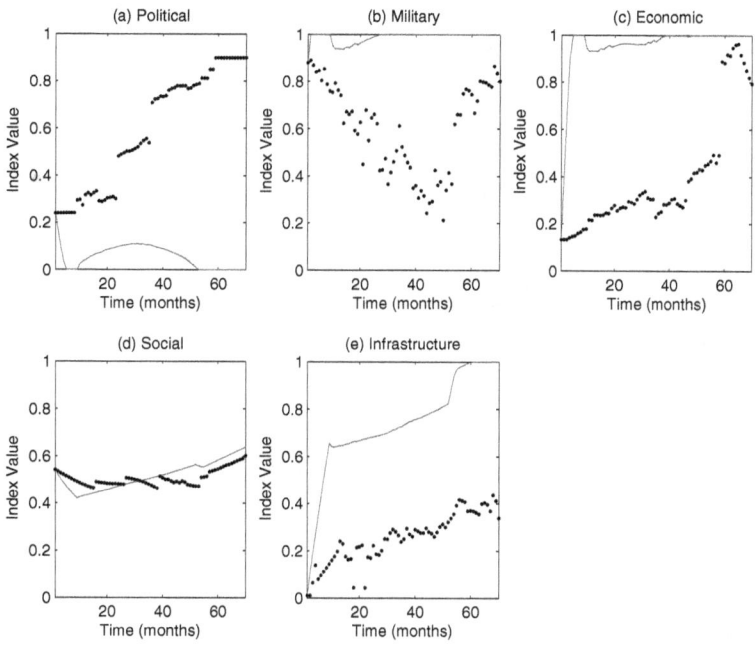

Figure 29. Run 8 States for Chapter IV

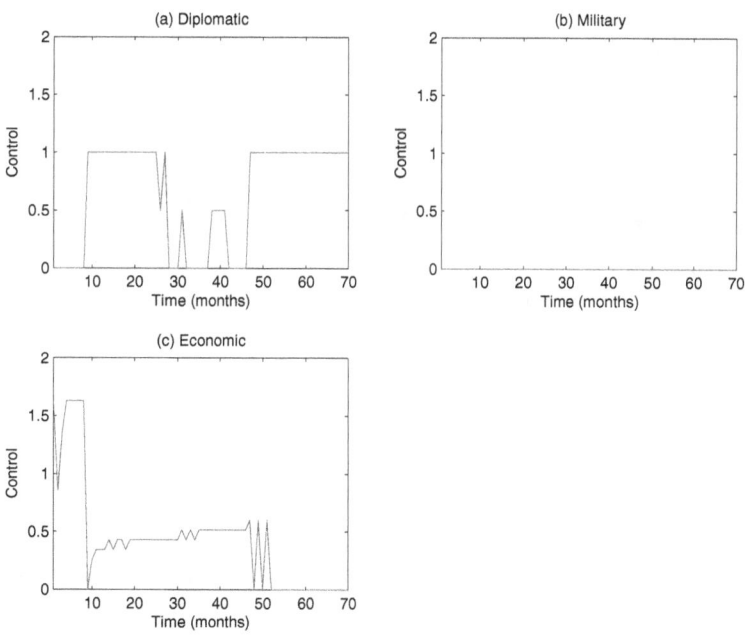

Figure 30. Run 8 Controls for Chapter IV

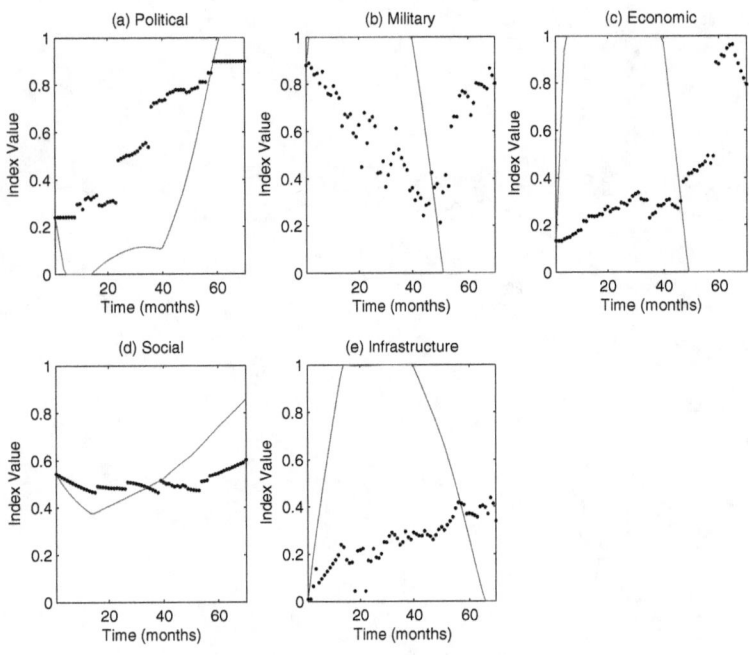

Figure 31. Run 9 States for Chapter IV

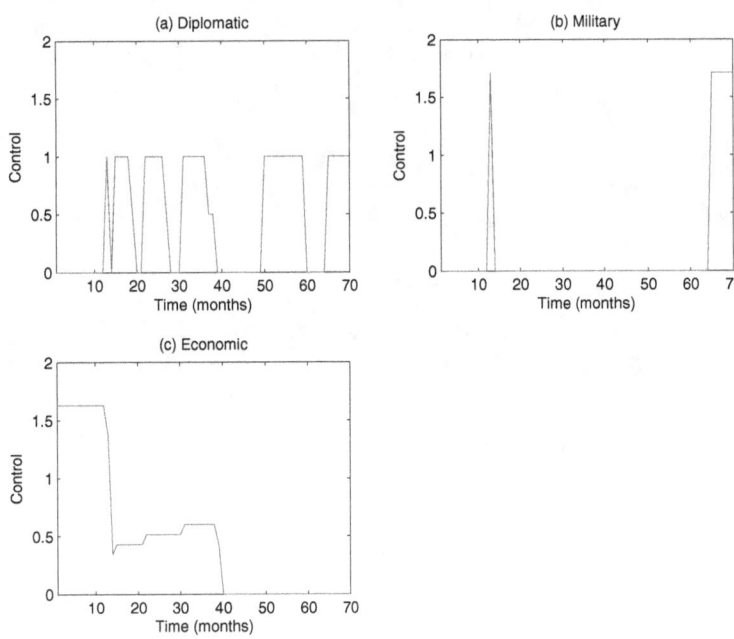

Figure 32. Run 9 Controls for Chapter IV

109

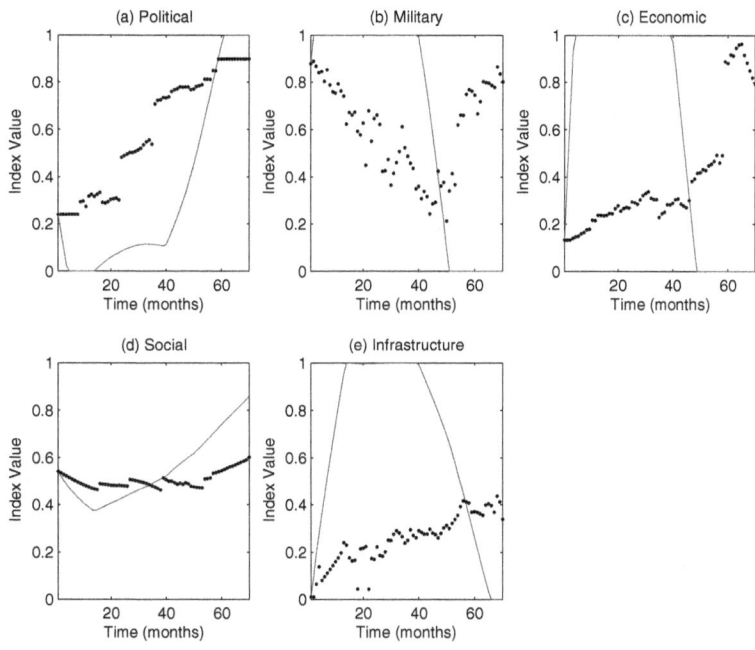

Figure 33. Run 10 States for Chapter IV

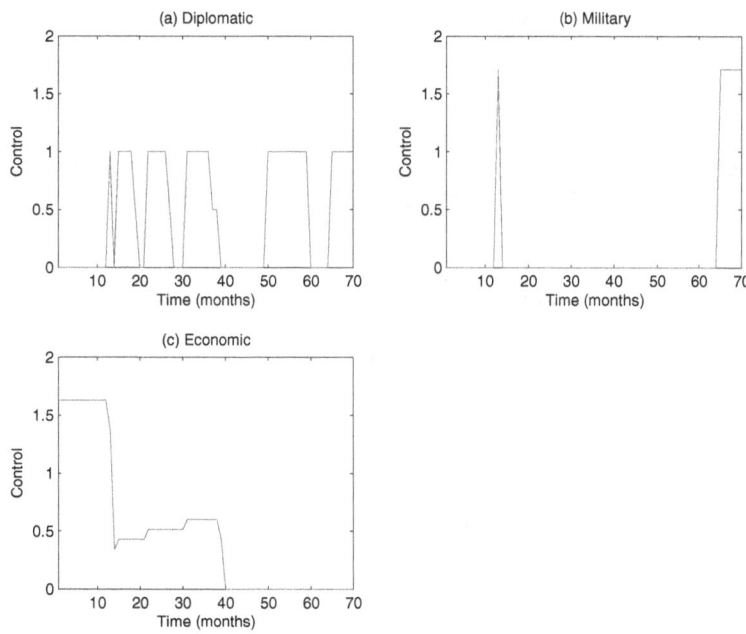

Figure 34. Run 10 Controls for Chapter IV

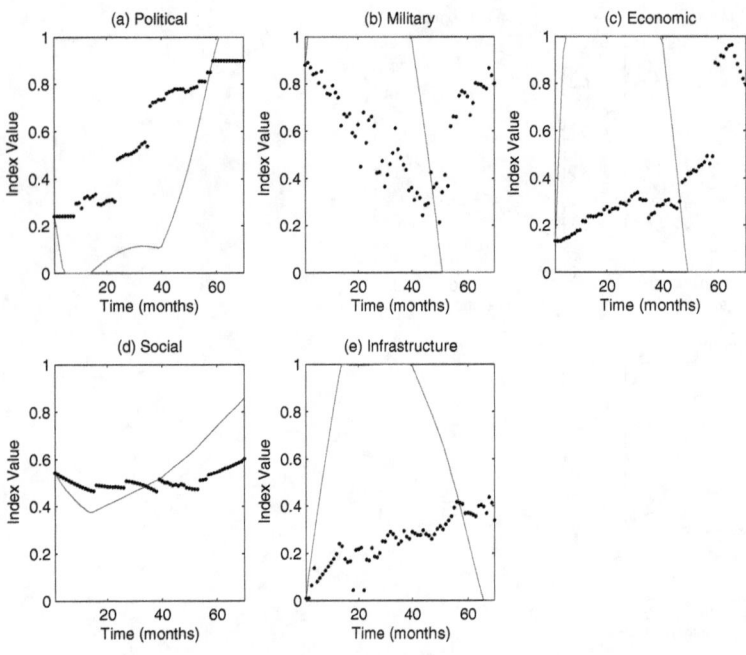

Figure 35. Run 11 States for Chapter IV

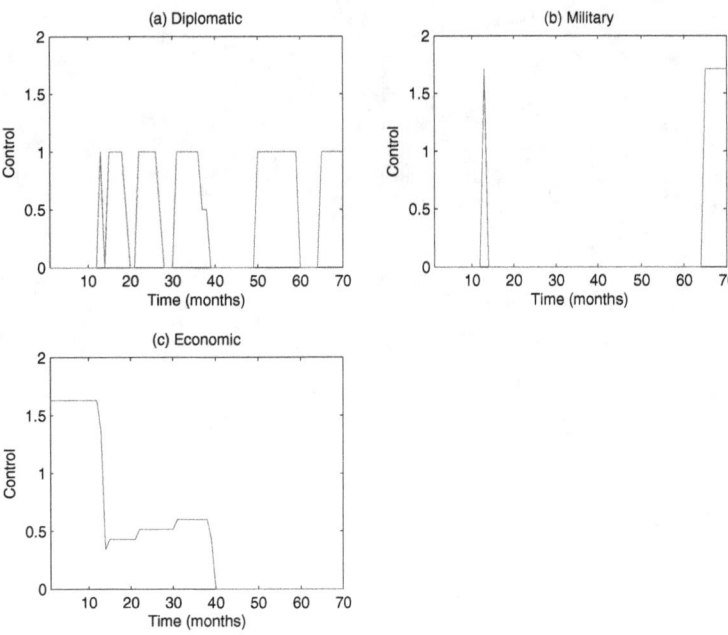

Figure 36. Run 11 Controls for Chapter IV

111

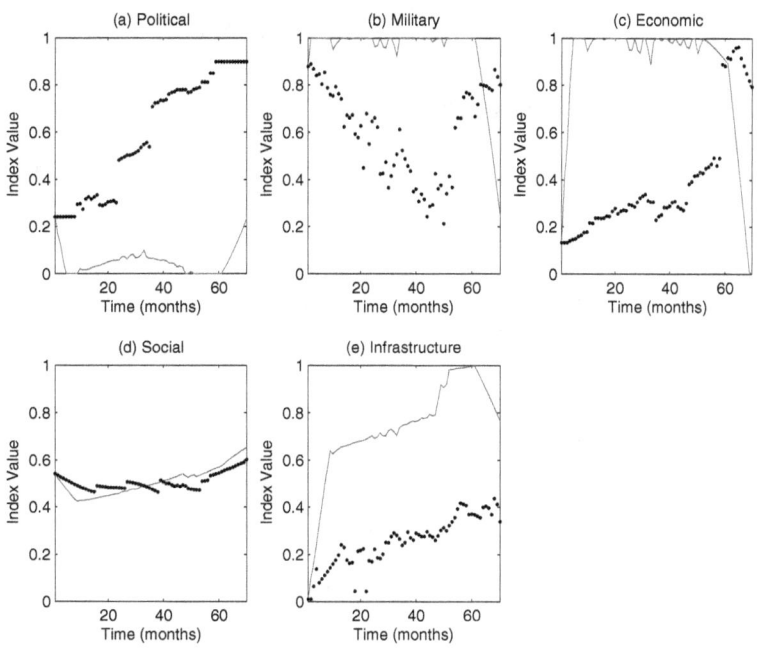

Figure 37. Run 12 States for Chapter IV

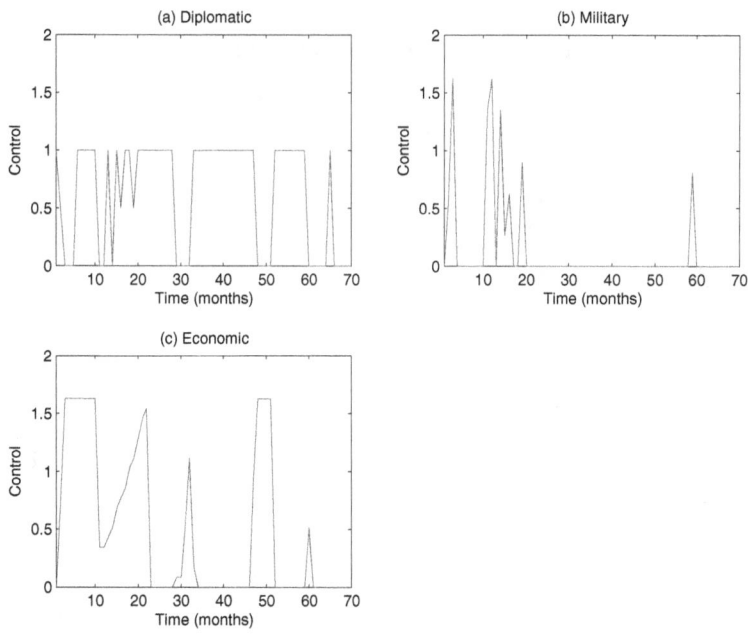

Figure 38. Run 12 Controls for Chapter IV

112

Appendix D. State and Control Plots for Chapter V with Actual Violence

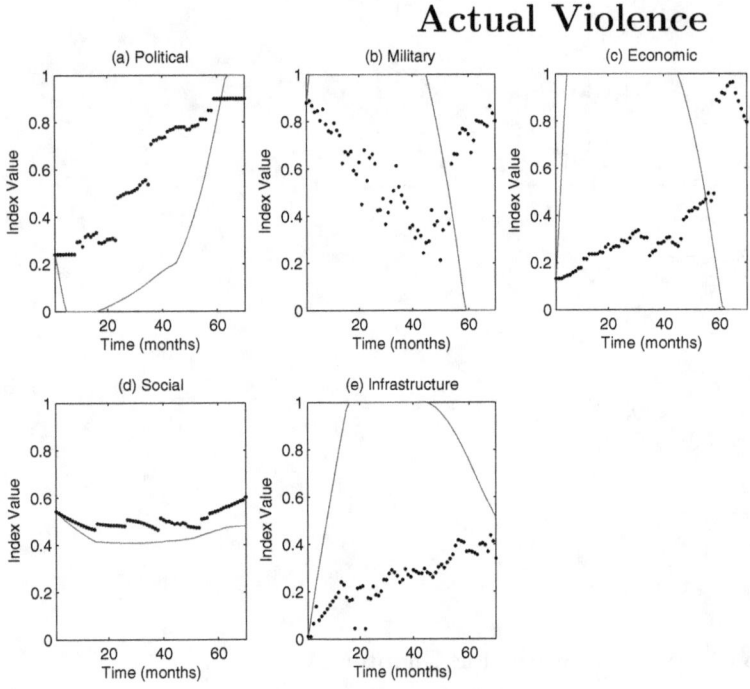

Figure 39. Run 1 States for Chapter V with Actual Violence

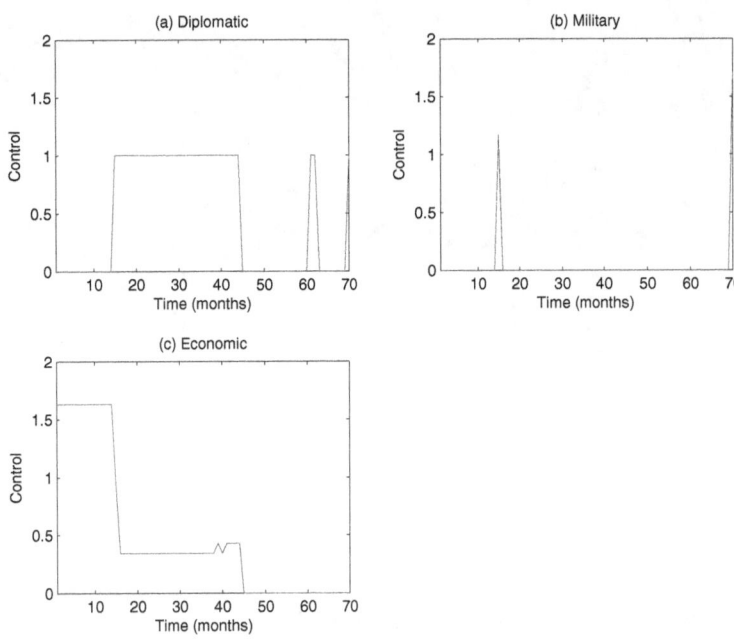

Figure 40. Run 1 Controls for Chapter V with Actual Violence

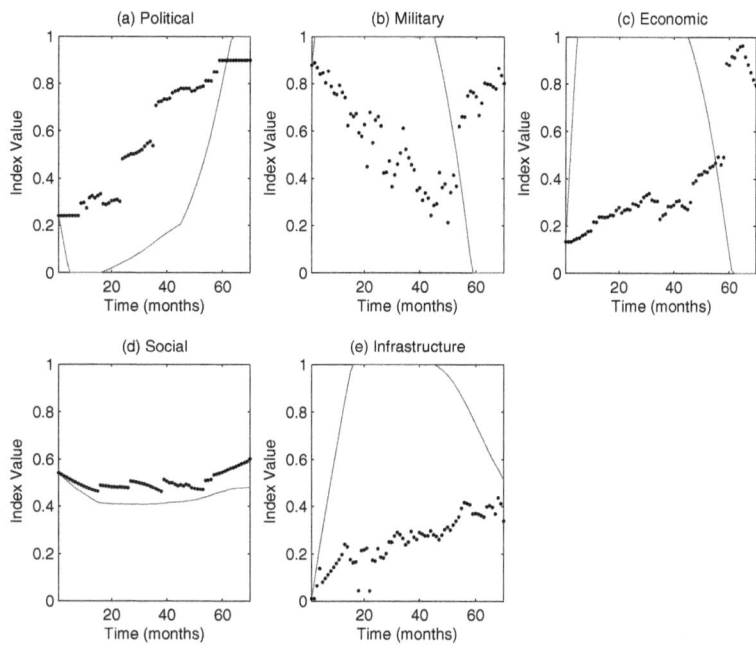

Figure 41. Run 2 States for Chapter V with Actual Violence

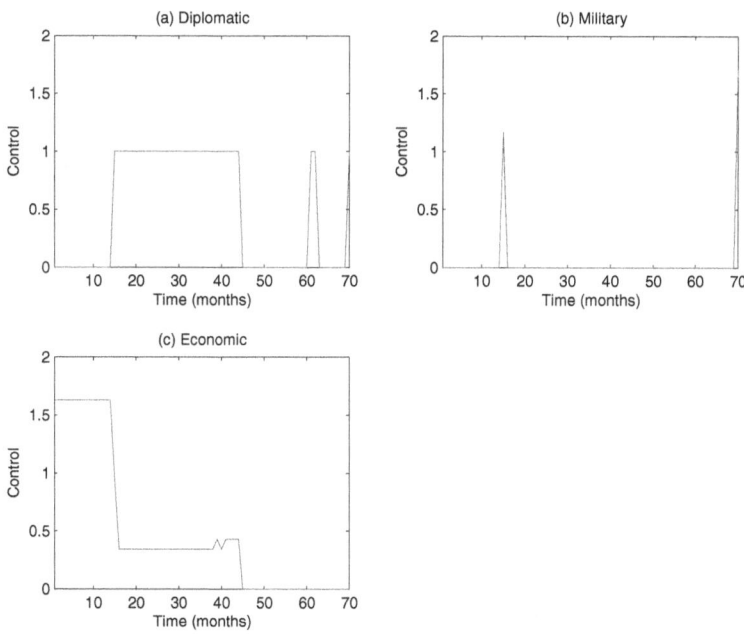

Figure 42. Run 2 Controls for Chapter V with Actual Violence

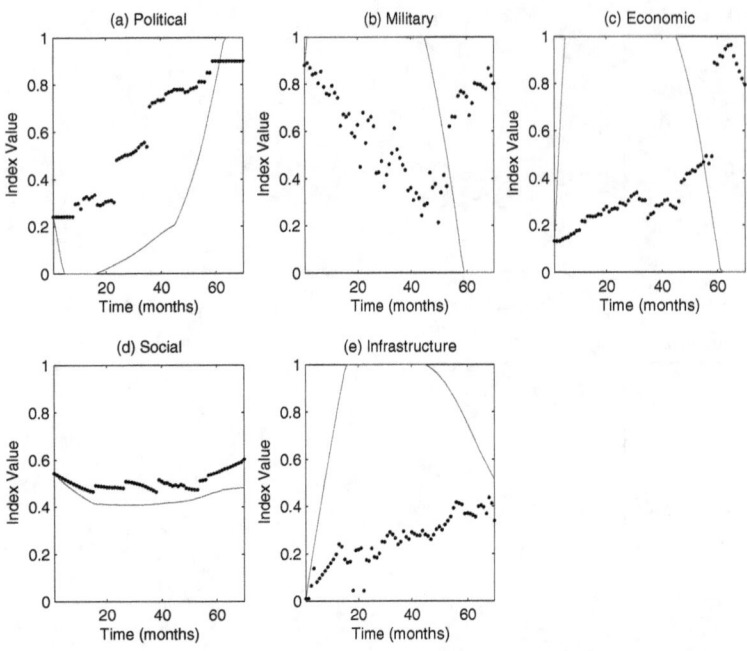

Figure 43. Run 3 States for Chapter V with Actual Violence

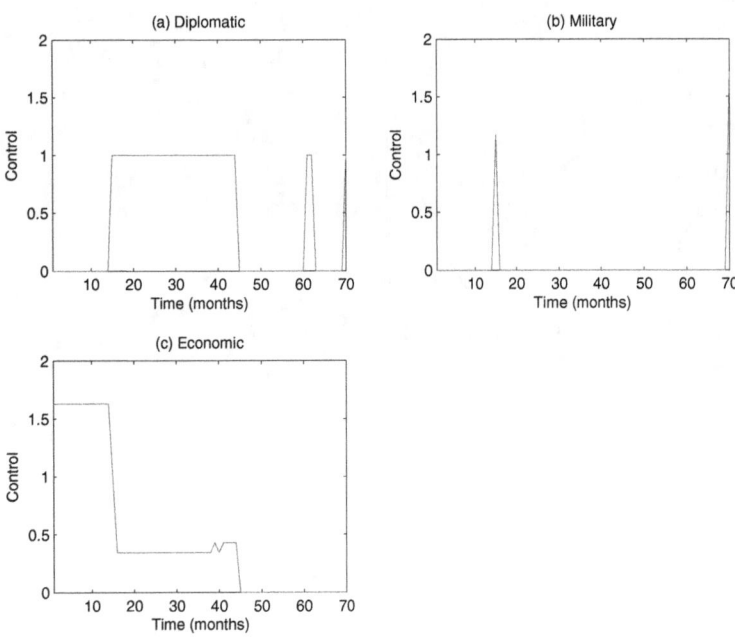

Figure 44. Run 3 Controls for Chapter V with Actual Violence

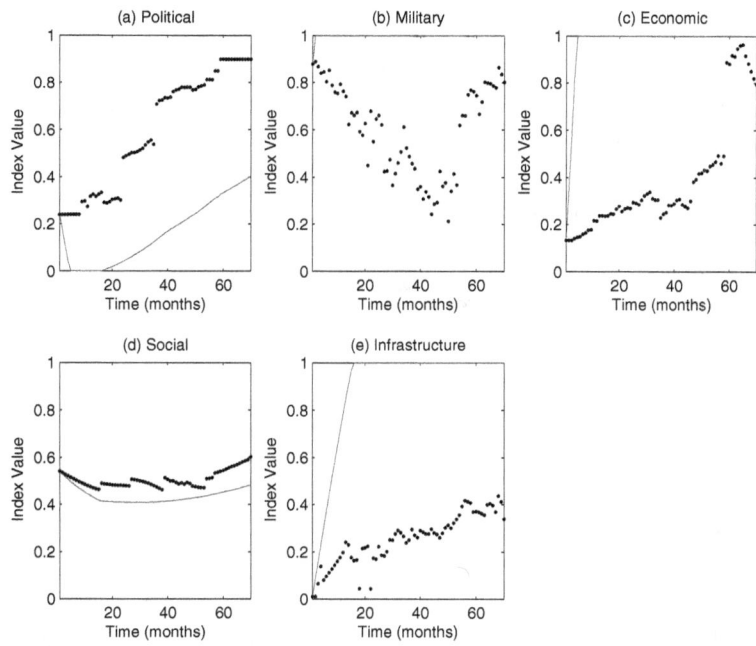

Figure 45. Run 4 States for Chapter V with Actual Violence

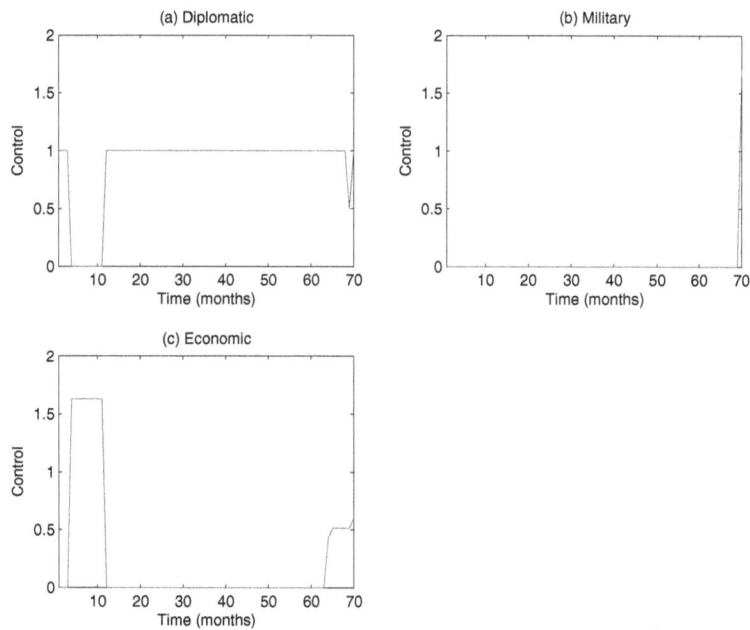

Figure 46. Run 4 Controls for Chapter V with Actual Violence

116

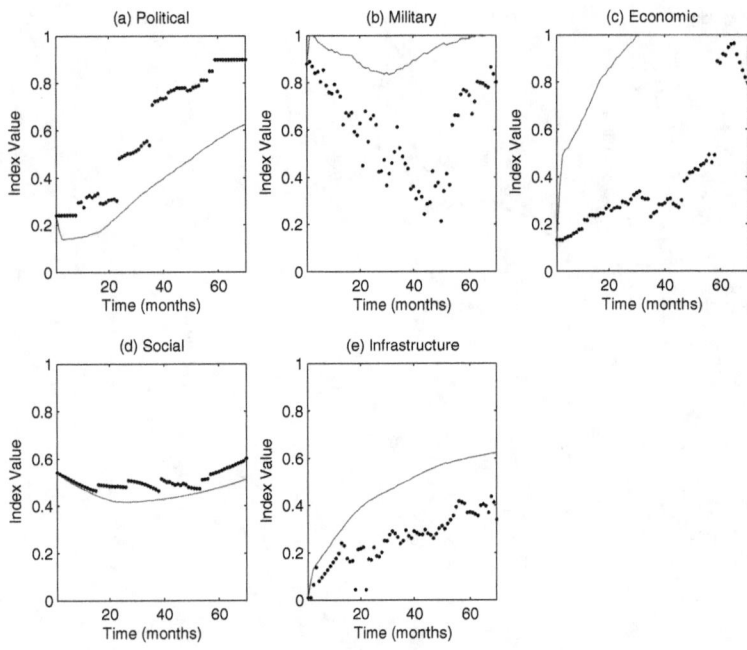

Figure 47. Run 5 States for Chapter V with Actual Violence

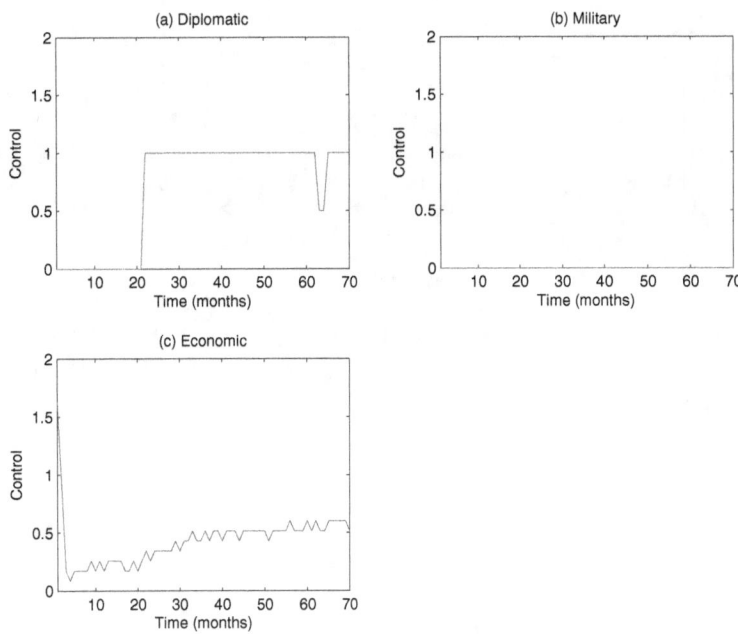

Figure 48. Run 5 Controls for Chapter V with Actual Violence

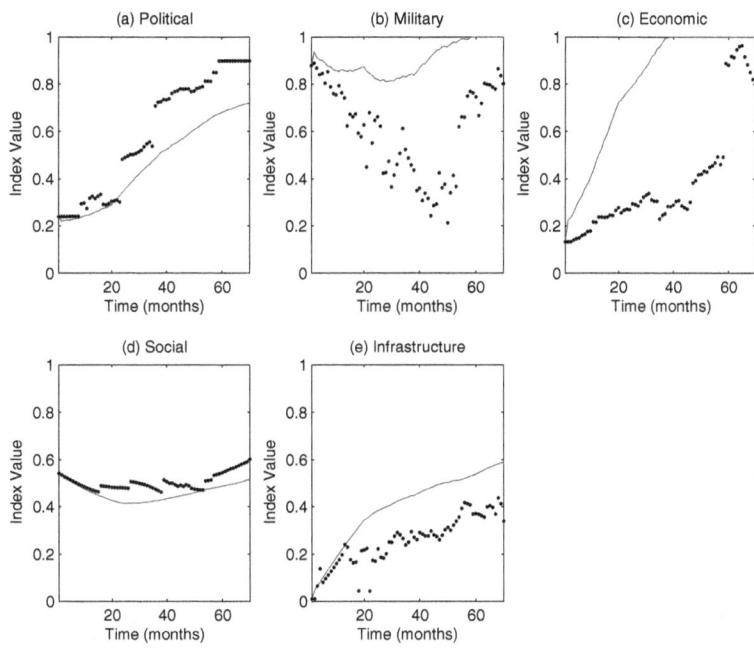

Figure 49. Run 6 States for Chapter V with Actual Violence

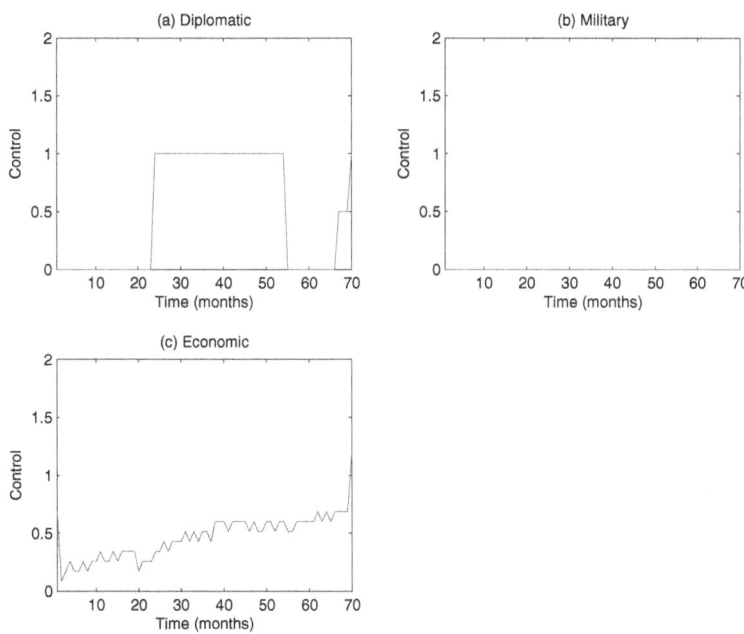

Figure 50. Run 6 Controls for Chapter V with Actual Violence

118

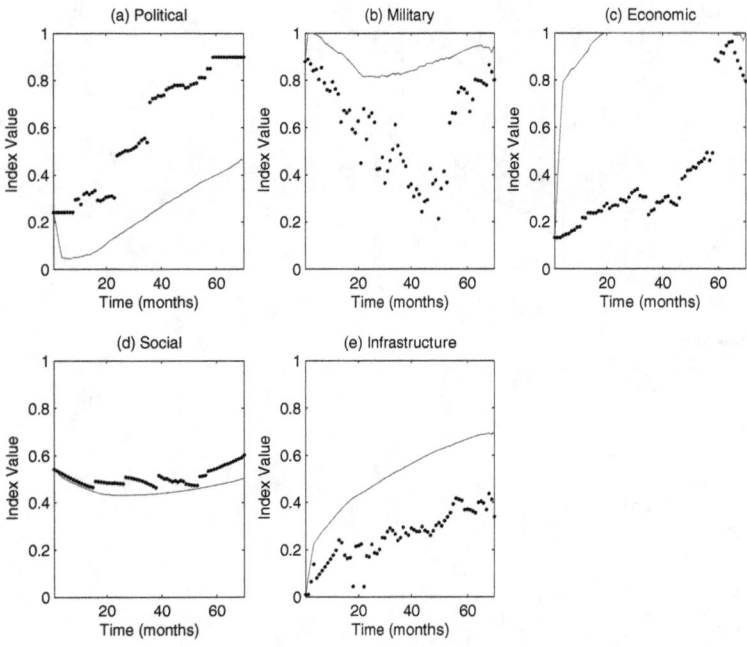

Figure 51. Run 7 States for Chapter V with Actual Violence

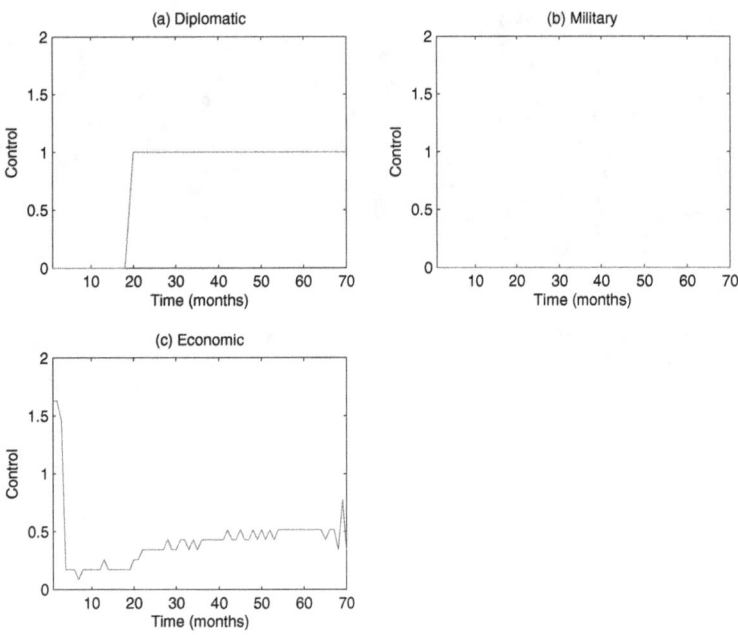

Figure 52. Run 7 Controls for Chapter V with Actual Violence

119

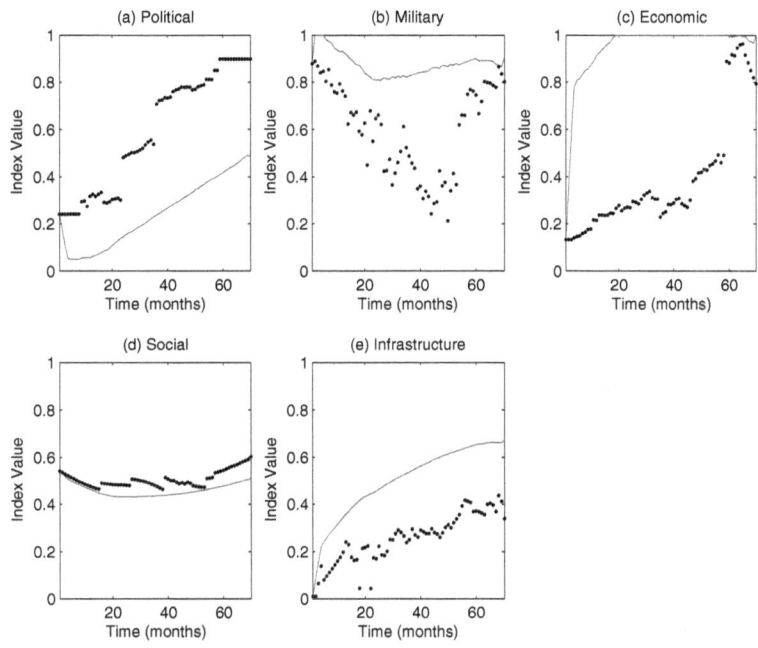

Figure 53. Run 8 States for Chapter V with Actual Violence

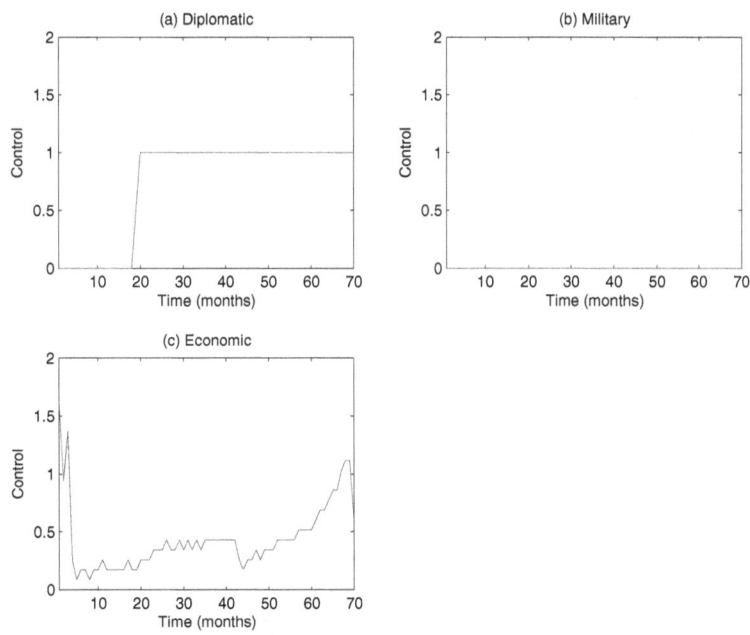

Figure 54. Run 8 Controls for Chapter V with Actual Violence

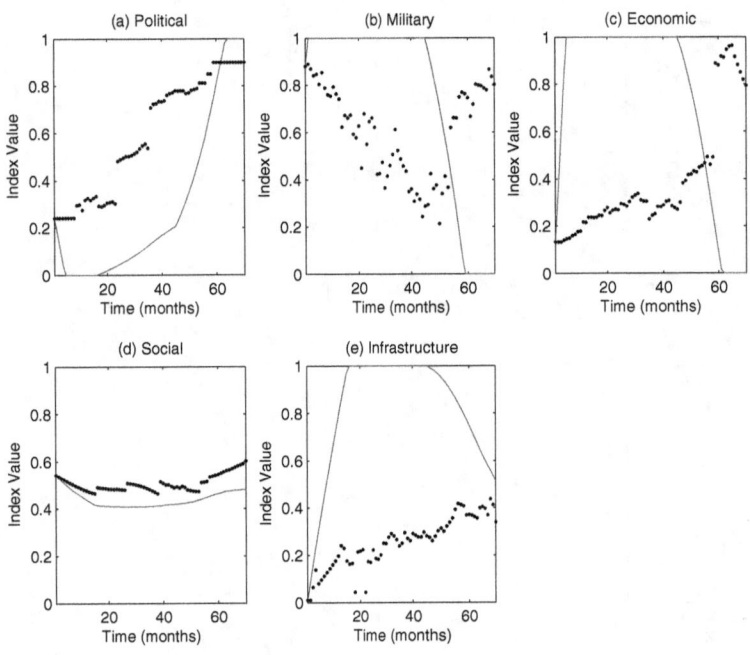

Figure 55. Run 9 States for Chapter V with Actual Violence

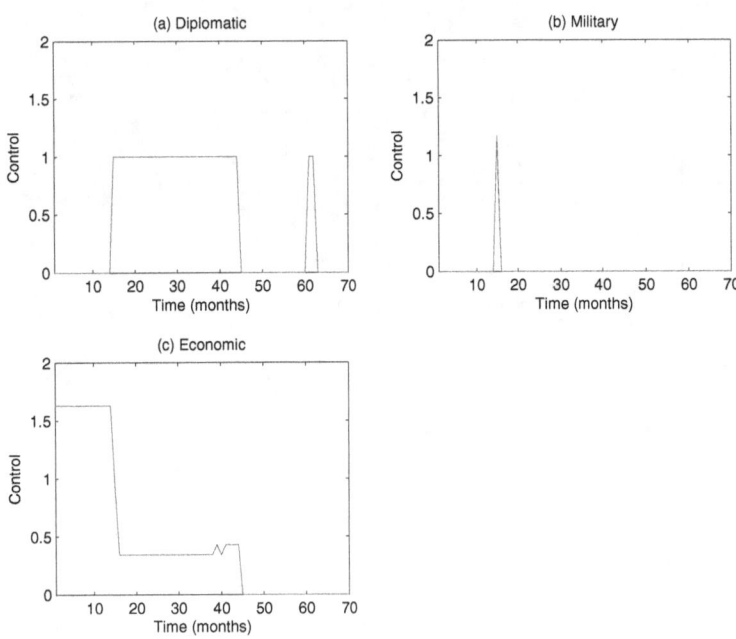

Figure 56. Run 9 Controls for Chapter V with Actual Violence

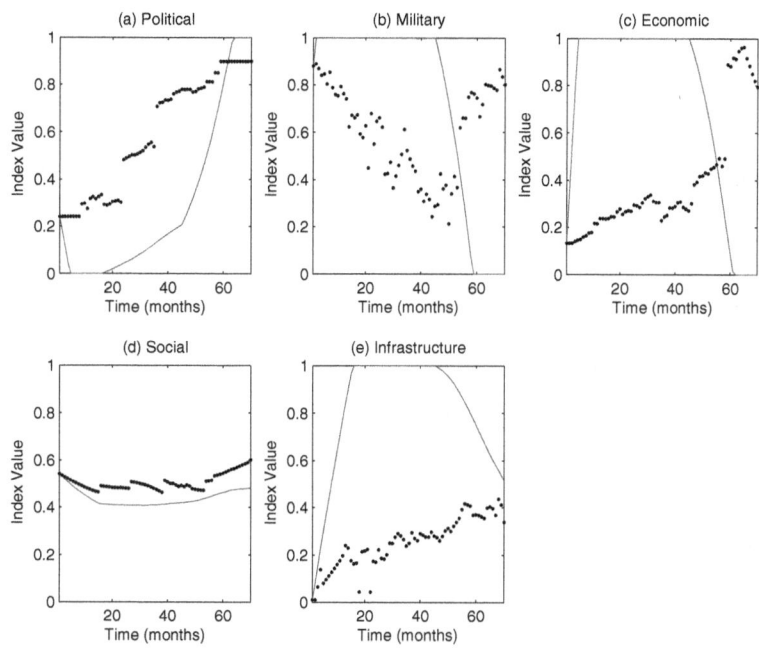

Figure 57. Run 10 States for Chapter V with Actual Violence

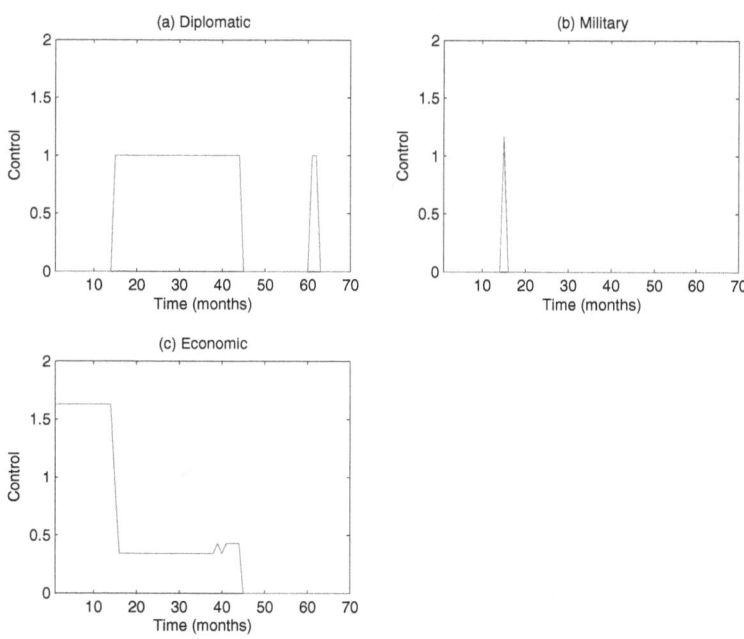

Figure 58. Run 10 Controls for Chapter V with Actual Violence

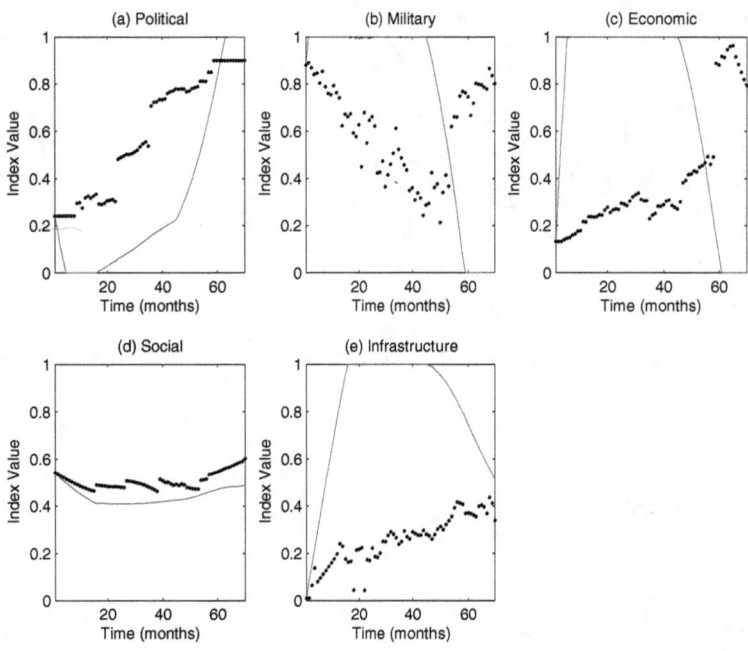

Figure 59. Run 11 States for Chapter V with Actual Violence

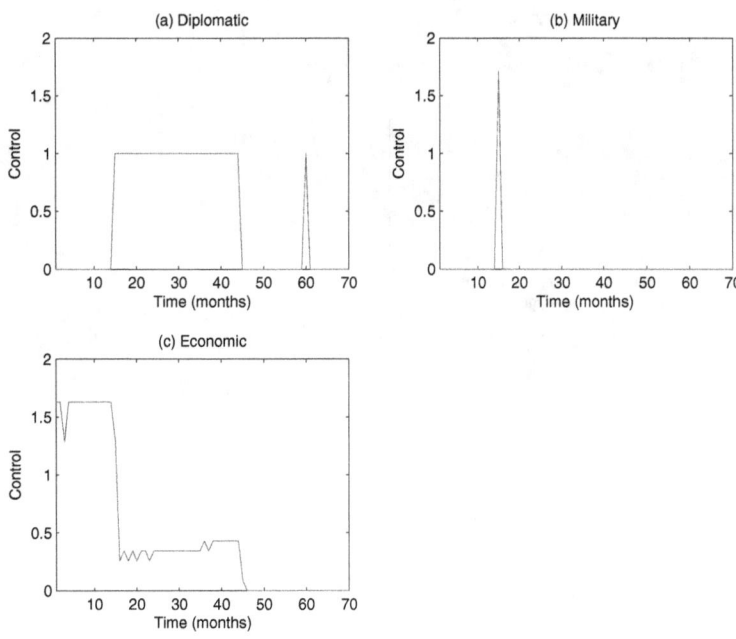

Figure 60. Run 11 Controls for Chapter V with Actual Violence

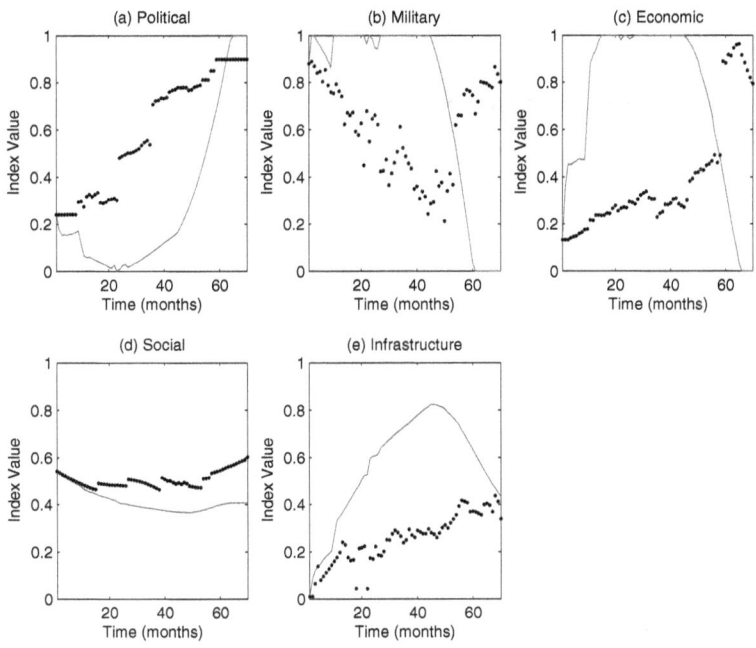

Figure 61. Run 12 States for Chapter V with Actual Violence

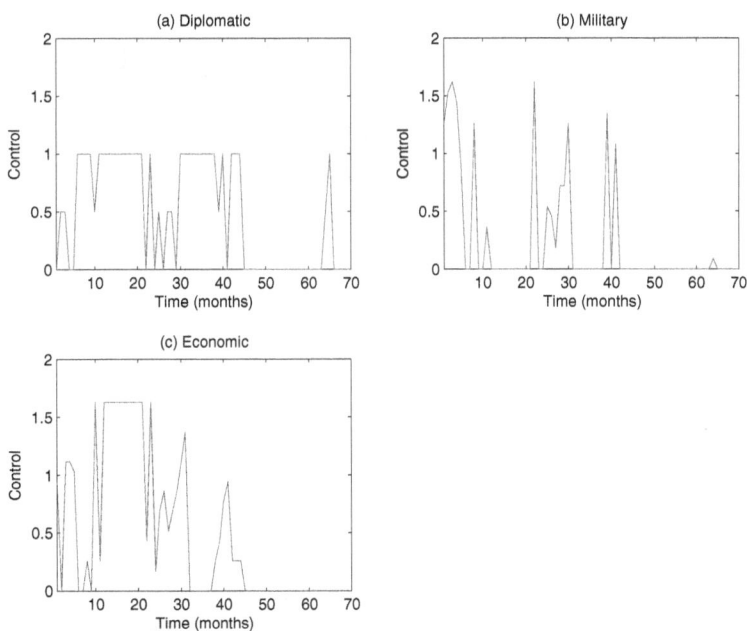

Figure 62. Run 12 Controls for Chapter V with Actual Violence

124

Appendix E. State and Control Plots for Chapter V with Calculated Violence

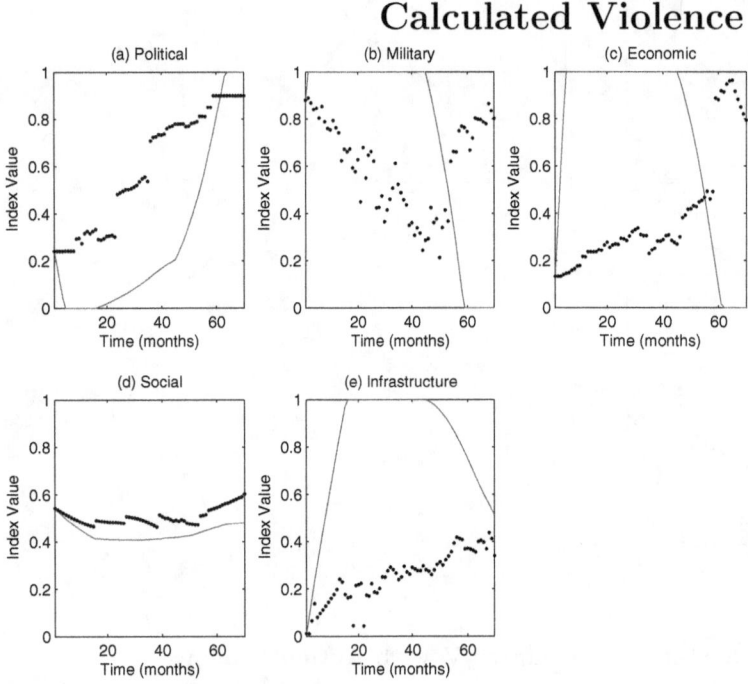

Figure 63. Run 1 States for Chapter V with Calculated Violence

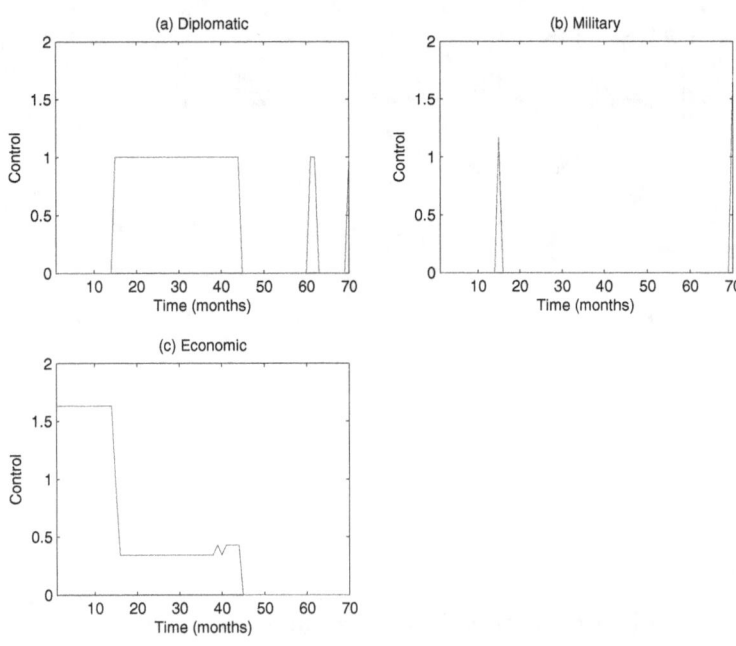

Figure 64. Run 1 Controls for Chapter V with Calculated Violence

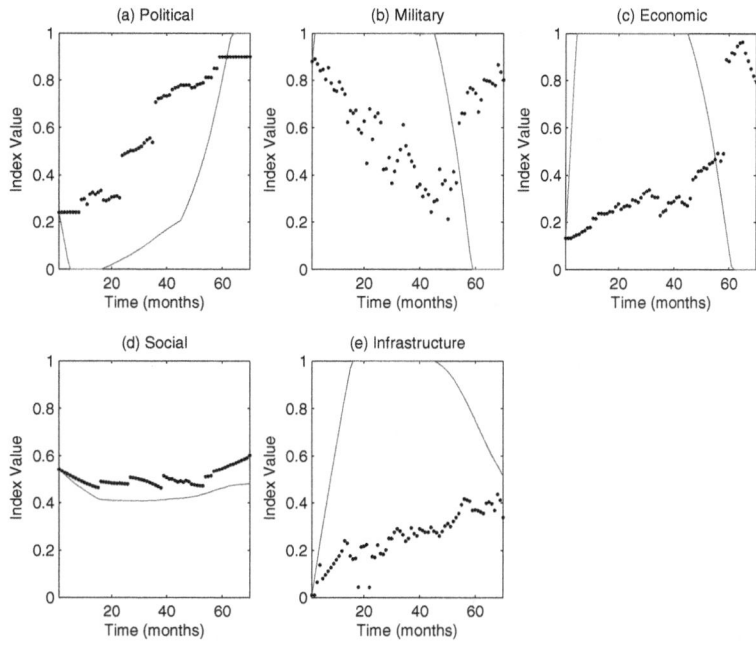

Figure 65. Run 2 States for Chapter V with Calculated Violence

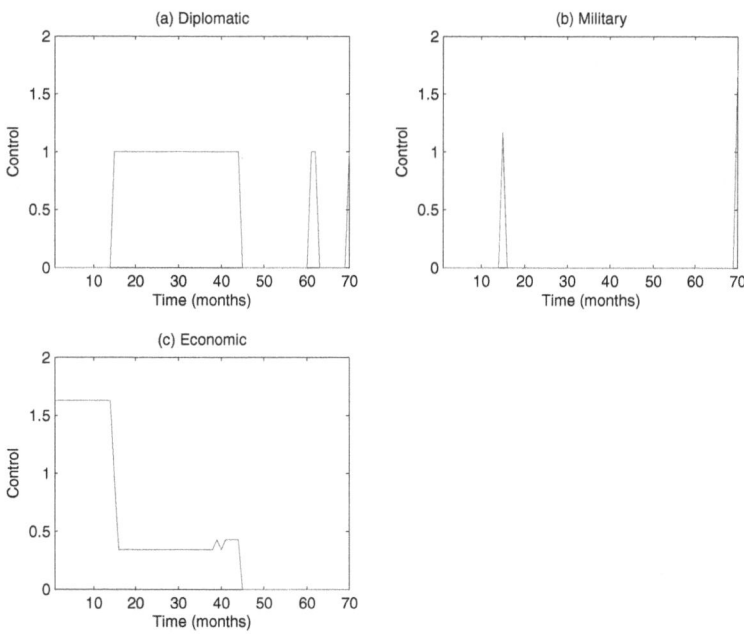

Figure 66. Run 2 Controls for Chapter V with Calculated Violence

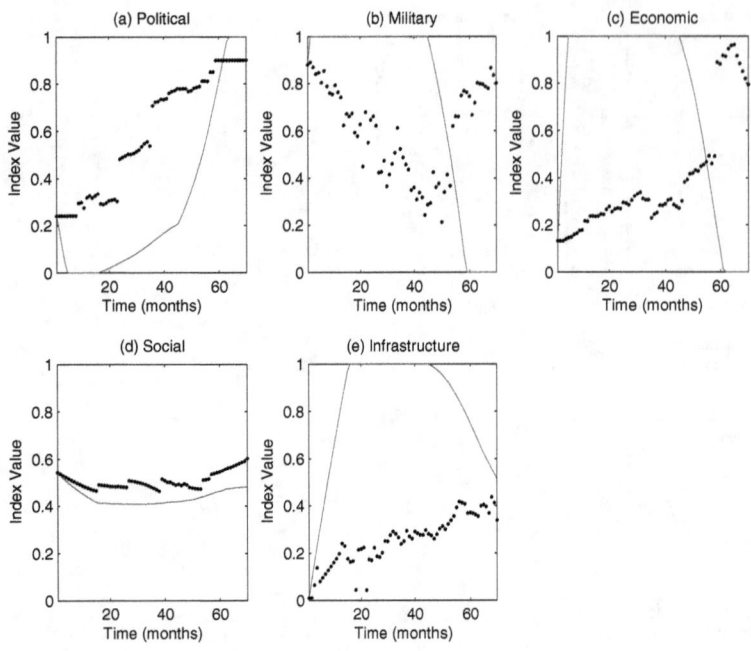

Figure 67. Run 3 States for Chapter V with Calculated Violence

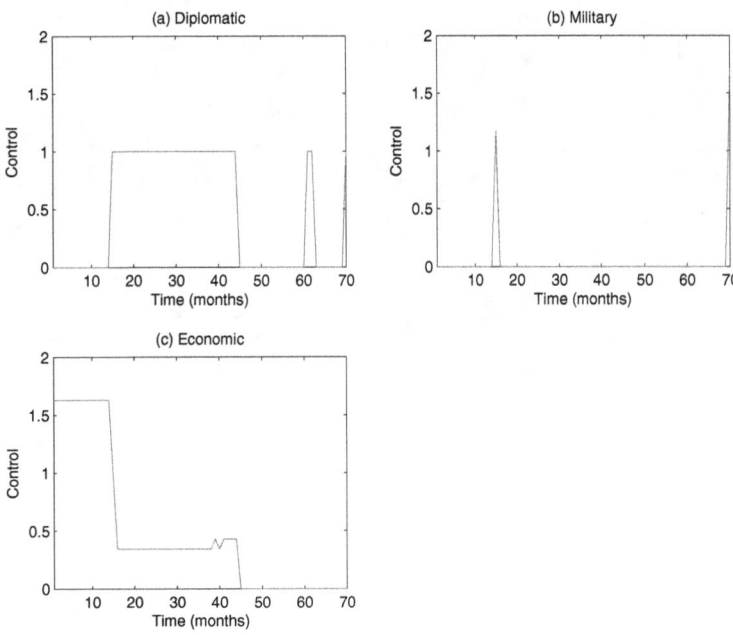

Figure 68. Run 3 Controls for Chapter V with Calculated Violence

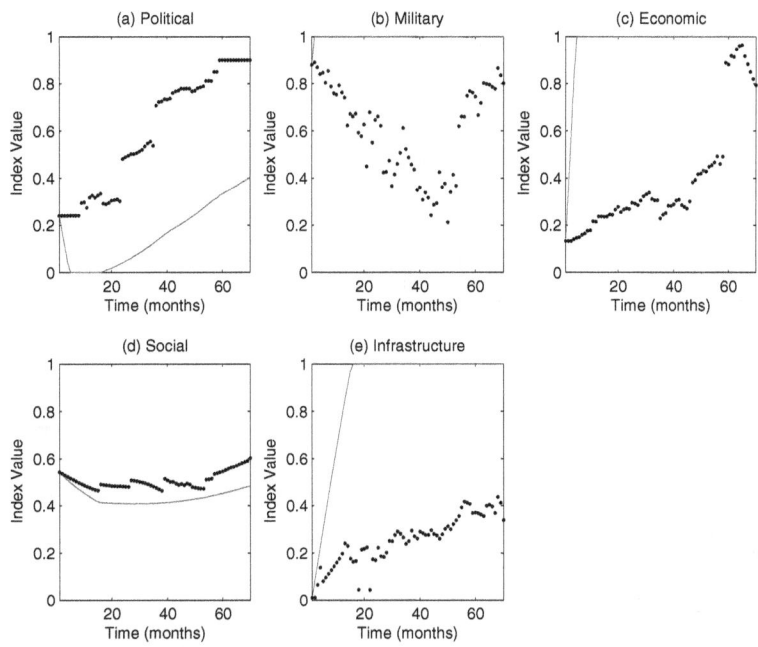

Figure 69. Run 4 States for Chapter V with Calculated Violence

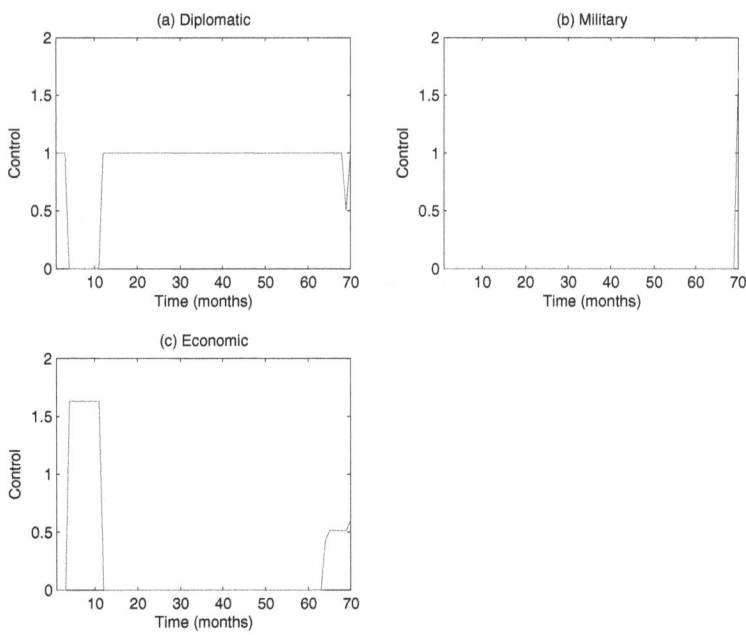

Figure 70. Run 4 Controls for Chapter V with Calculated Violence

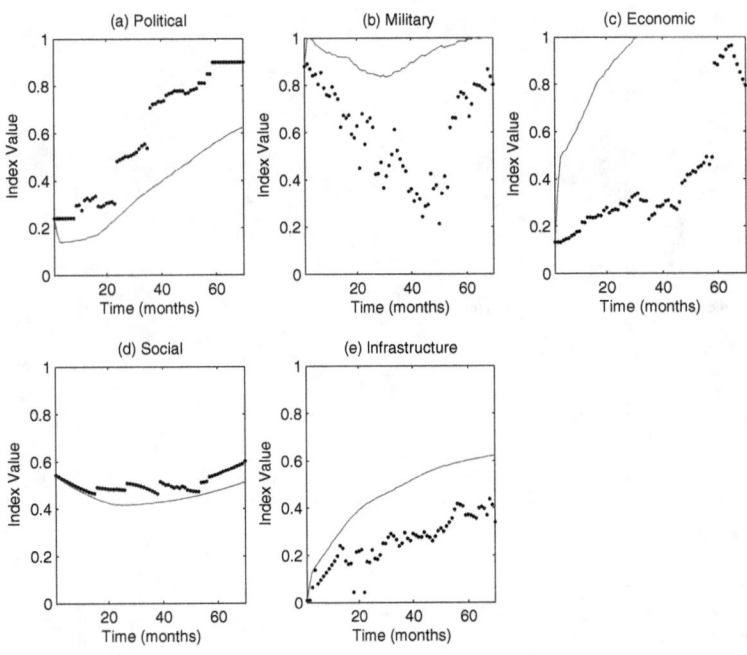

Figure 71. Run 5 States for Chapter V with Calculated Violence

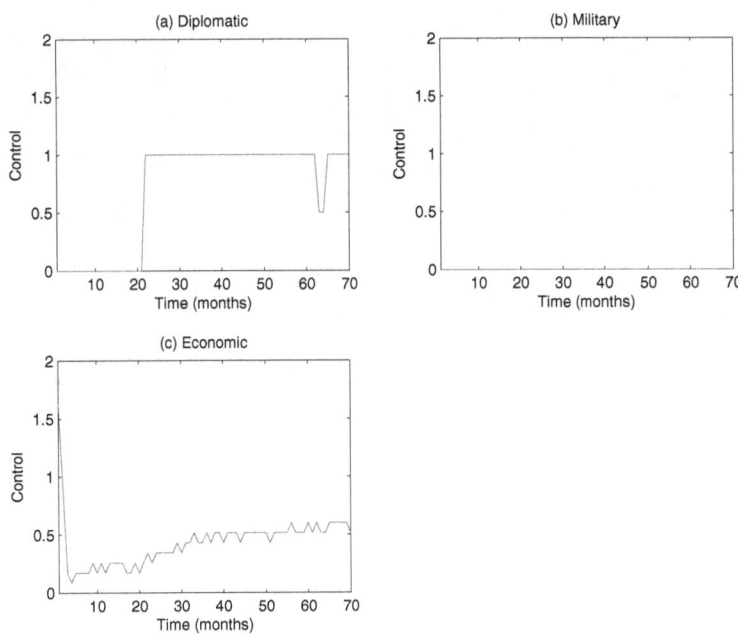

Figure 72. Run 5 Controls for Chapter V with Calculated Violence

129

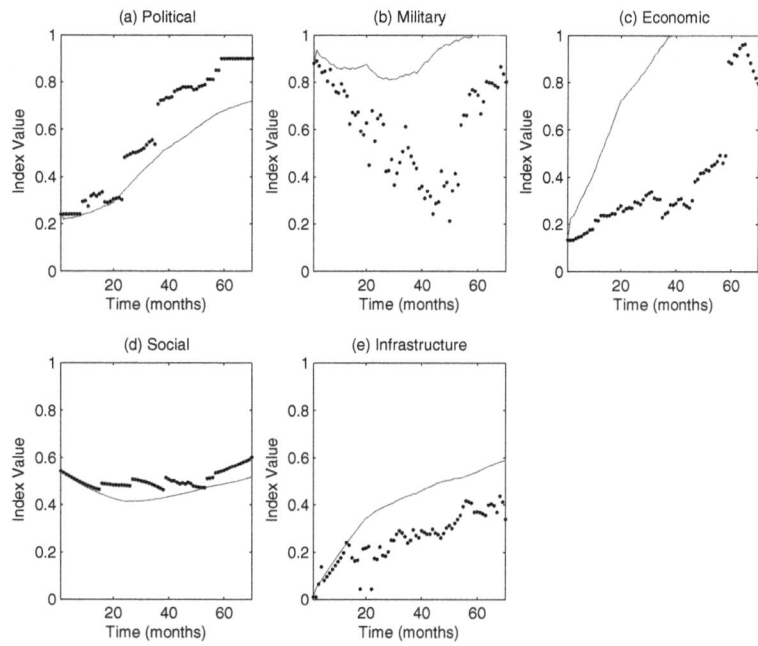

Figure 73. Run 6 States for Chapter V with Calculated Violence

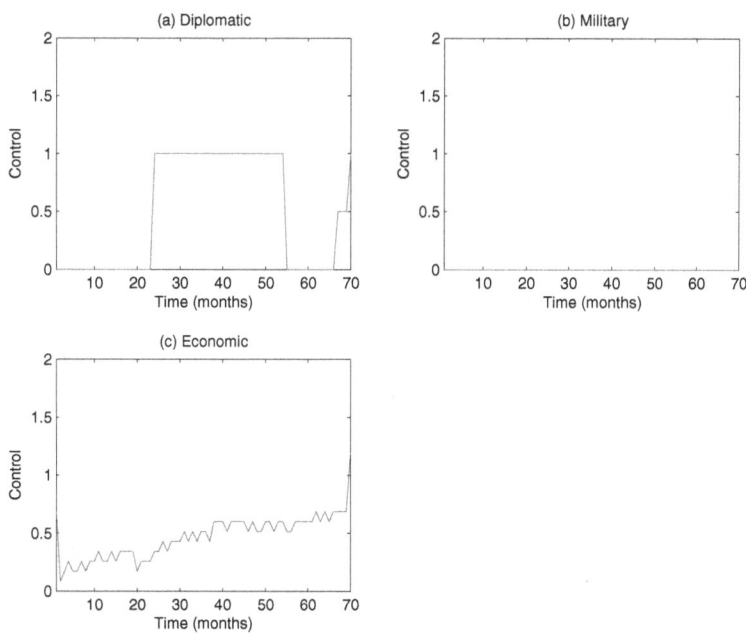

Figure 74. Run 6 Controls for Chapter V with Calculated Violence

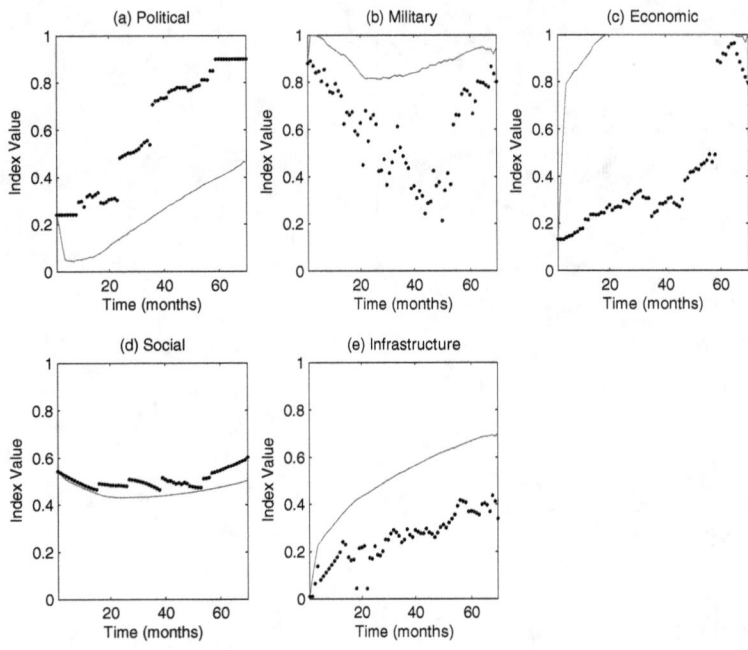

Figure 75. Run 7 States for Chapter V with Calculated Violence

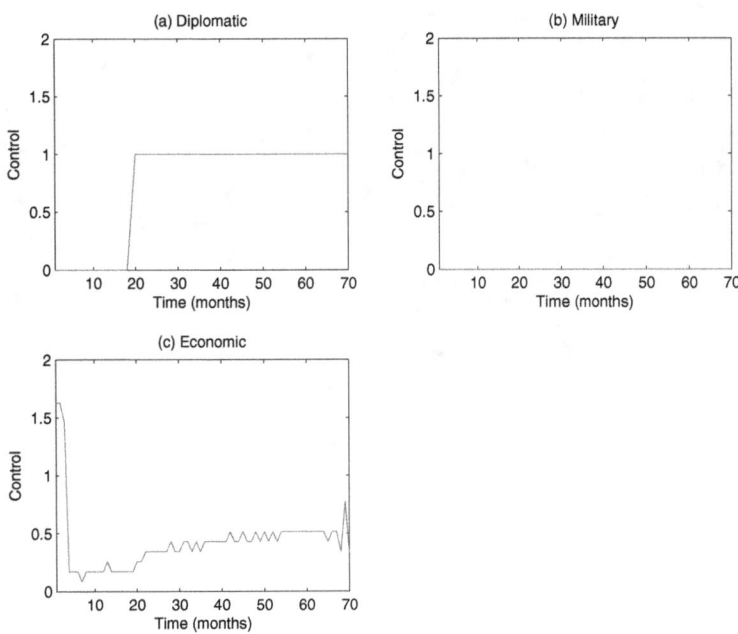

Figure 76. Run 7 Controls for Chapter V with Calculated Violence

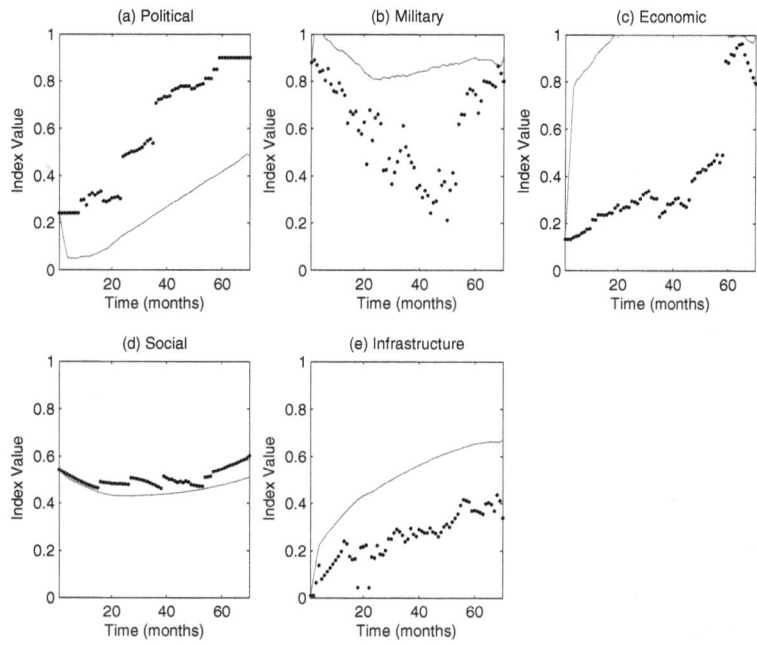

Figure 77. Run 8 States for Chapter V with Calculated Violence

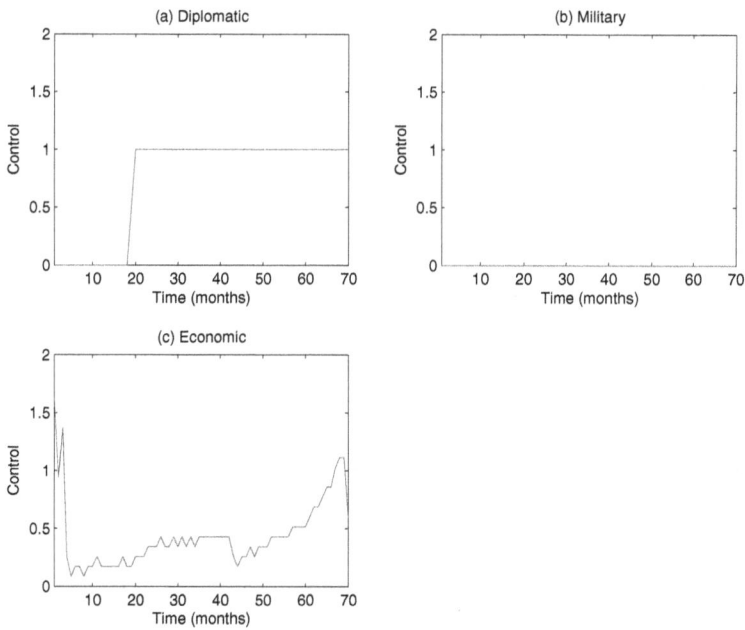

Figure 78. Run 8 Controls for Chapter V with Calculated Violence

132

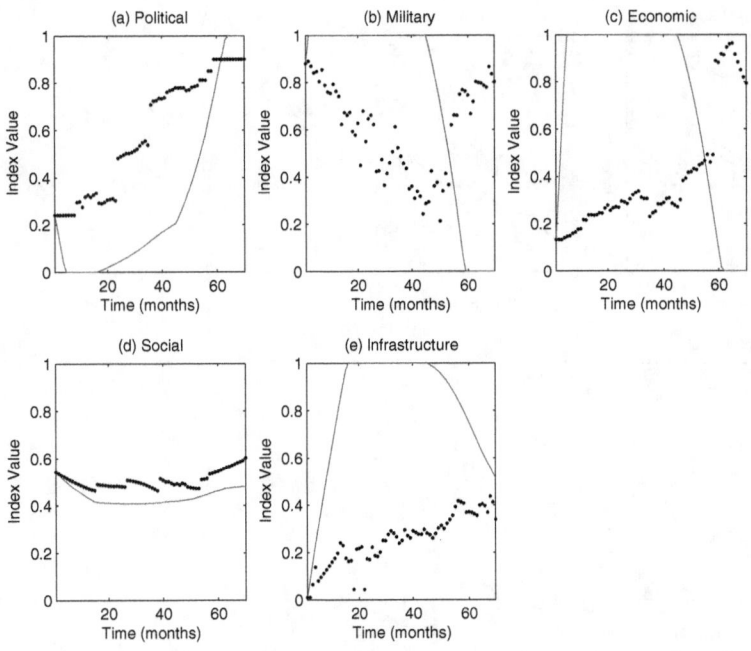

Figure 79. Run 9 States for Chapter V with Calculated Violence

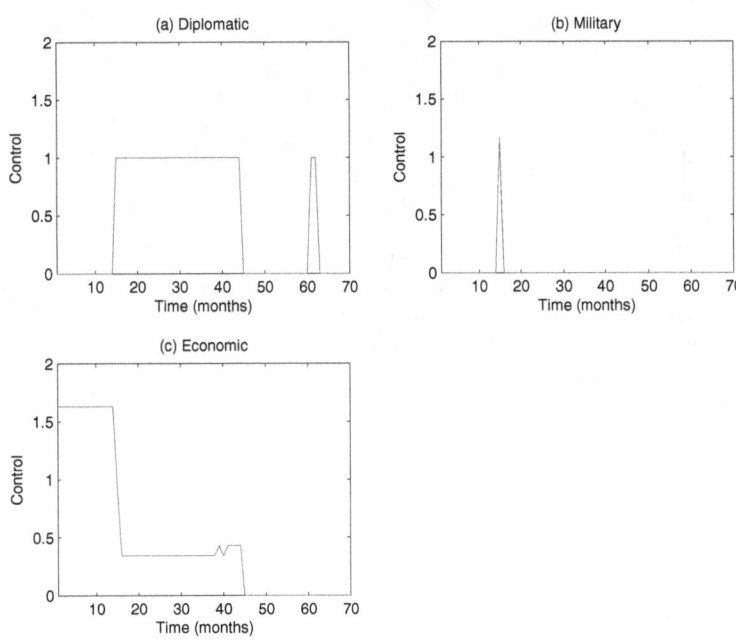

Figure 80. Run 9 Controls for Chapter V with Calculated Violence

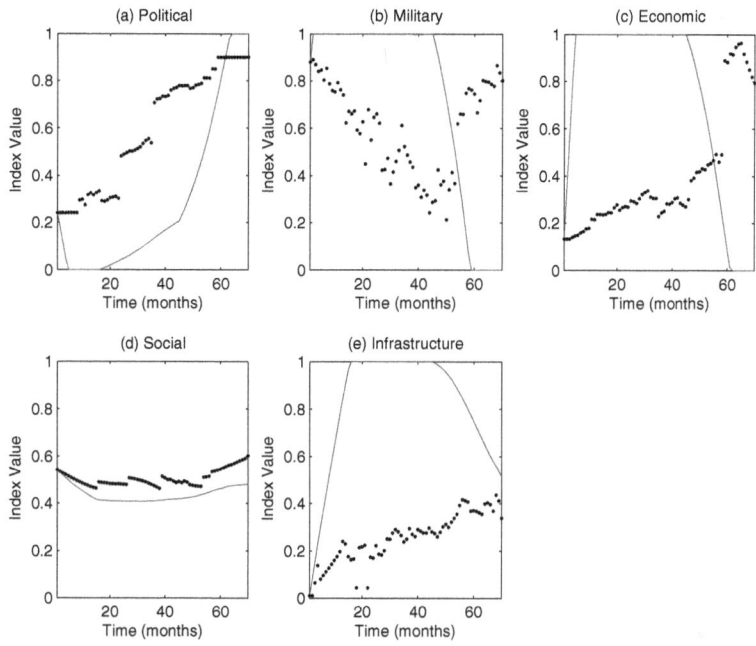

Figure 81. Run 10 States for Chapter V with Calculated Violence

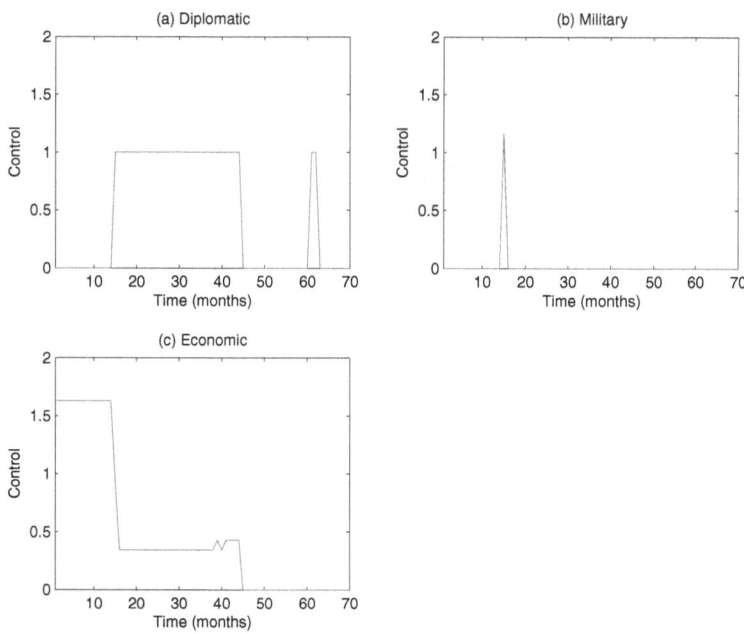

Figure 82. Run 10 Controls for Chapter V with Calculated Violence

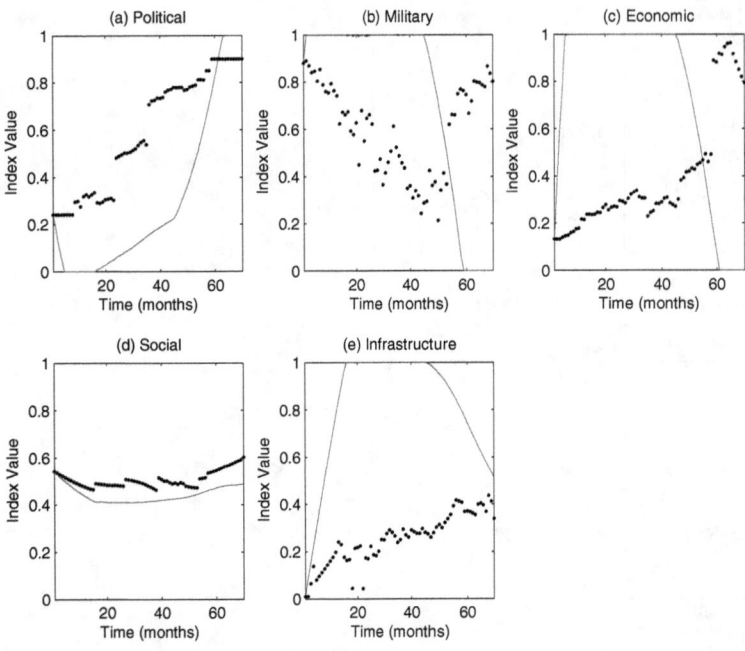

Figure 83. Run 11 States for Chapter V with Calculated Violence

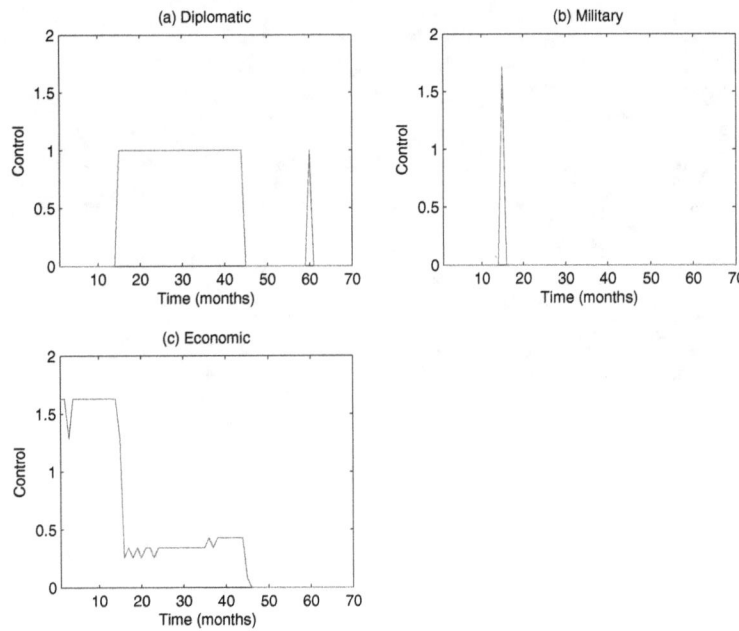

Figure 84. Run 11 Controls for Chapter V with Calculated Violence

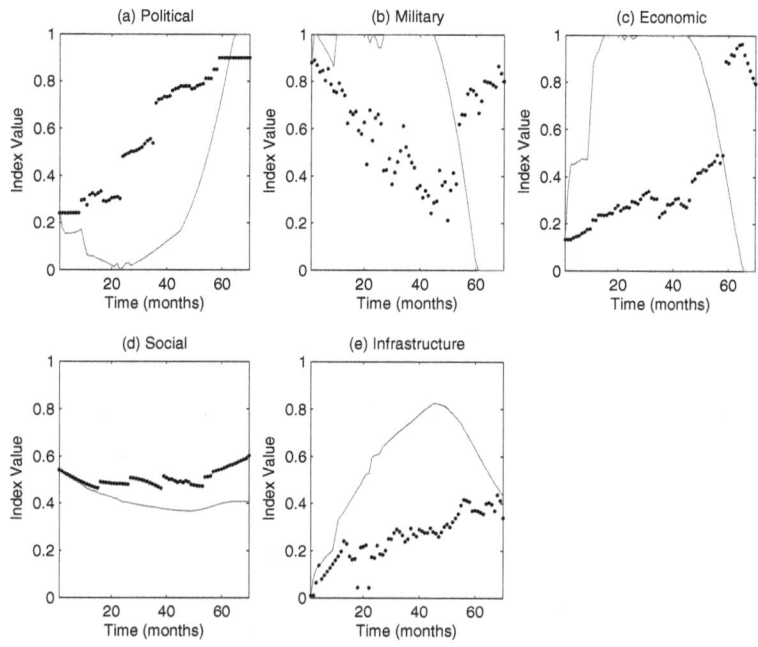

Figure 85. Run 12 States for Chapter V with Calculated Violence

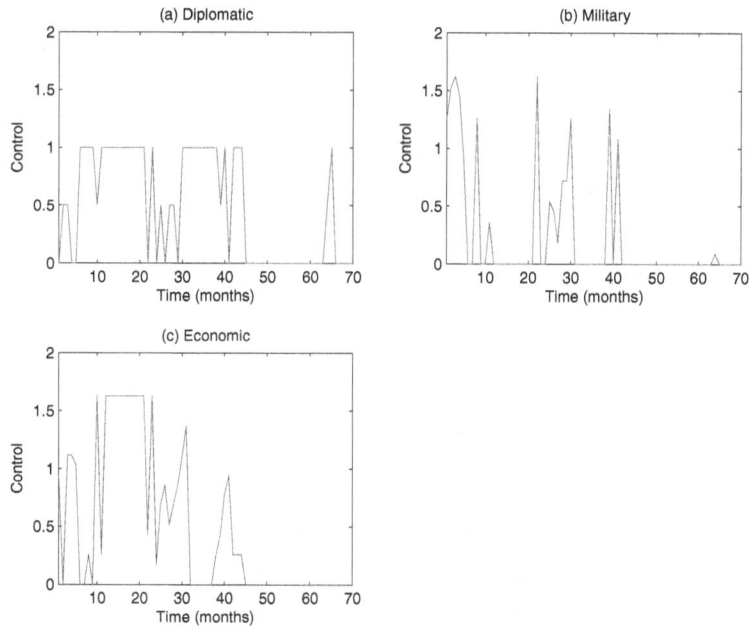

Figure 86. Run 12 Controls for Chapter V with Calculated Violence

Bibliography

[1] Abdollahian, Mark, Brice L. Nicholson, Matthew Nickens, and Michael Baranick. "A Formal Model of Stabilization and Reconstruction Operations", *Military Operations Research*, 14(3):5–30, 2009.

[2] Abdolliahian, M., M. Baranick, B. Efird, and J. Kugler. *Senturion: A Predictive Political Simulation Model.* Technical report, Center for Technology and National Security Policy, National Defense University, 2006.

[3] Adler, R.M. "A Dynamic Social Network Software Platform for Counter-Terrorism Decision Support". *Intelligence and Security Informatics, 2007 IEEE*, 47–54. 2007.

[4] Arney, David Chris and Kristin Arney. "Modeling insurgency, counter-insurgency, and coalition strategies and operations", *The Journal of Defense Modeling and Simulation: Applications, Methodology, Technology*, 10(1):57–73, 2013.

[5] Atkinson, M P, A Gutfraind, and M Kress. "When do armed revolts succeed: lessons from Lanchester theory", *Journal of the Operational Research Society*, 63(10):1363–1373, Oct 2012.

[6] Bandura, Romina. "A Survey of Composite Indices Measuring Country Performance: 2008 Update", 2008.

[7] Bang, Sungwan. *Coalition Modeling in Humanitarian Assistance Operations.* Master's thesis, Air Force Institute of Technology, 2006.

[8] Bellman, R.E. and R.S. Roth. *The Bellman Continuum: A Collection of the Works of Richard E. Bellman.* World Scientific Publishing Company, Incorporated, 1986.

[9] Bernardoni, B. J., R. F. Deckro, and M. J. Robbins. "Using Social Network Analysis to Inform Stabilization Efforts", *Military Operations Research*, 18(4):37–60, 2013.

[10] Bertazzi, Luca, Adamo Bosco, Francesca Guerriero, and Demetrio Lagan. "A stochastic inventory routing problem with stock-out", *Transportation Research Part C: Emerging Technologies*, 27(0):89 – 107, 2013.

[11] Bertsekas, Dimitri P. *Dynamic Programming and Optimal Control, Vol. I, 3rd Ed.* Belmont, MA: Athena Scientific, 2005.

[12] Bertsekas, Dimitri P. "Dynamic Programming and Suboptimal Control: A Survey from {ADP} to MPC*", *European Journal of Control*, 11(4-5):310 – 334, 2005.

[13] Bertsekas, Dimitri P. and David A. Castaon. "Rollout algorithms for stochastic scheduling problems", *Journal of Heuristics*, 89–108, 1999.

[14] Bertsekas, Dimitri P. and John N. Tsitsiklis. *Neuro-Dynamic Programming*, 1st edition). Athena Scientific, 1996.

[15] Bertsekas, Dimitri P., John N. Tsitsiklis, and Cynara Wu. "Rollout Algorithms for Combinatorial Optimization", *Journal of Heuristics*, 3(3):245–262, November 1997.

[16] Bertsekas, Dimitri.P. *Rollout Algorithms for Constrained Dynamic Programming*. LIDS 2646, Massachussetts Institute of Technology, 2005.

[17] Bertsekas, DimitriP. "Rollout Algorithms for Discrete Optimization: A Survey". Panos M. Pardalos, Ding-Zhu Du, and Ronald L. Graham (editors), *Handbook of Combinatorial Optimization*, 2989–3013. Springer New York, 2013. ISBN 978-1-4419-7996-4. URL http://dx.doi.org/10.1007/978-1-4419-7997-1_8.

[18] Blank, Karry, Carl E. Enomoto, Douglas Gegax, Thomas McGuckin, and Cade Simmons. "A Dynamic Model of Insurgency: The Case of the War in Iraq", *Peace Economics, Peace Science and Public Policy*, 14-2:1–26, 2008.

[19] Boccara, Nino. *Modeling Complex Systems*. New York: Springer, 2010.

[20] Body, Howard and Colin Marston. "The Peace Support Operations Model: Origins, Development, Philosophy and Use", *The Journal of Defense Modeling and Simulation: Applications, Methodology, Technology*, 8(2):69–77, 2011.

[21] Bonder, Seth. personal communication with Darryl Ahner, April 2008.

[22] Bracken, J. "Lanchester Models of the Ardennes Campaign", *Naval Research Logistics*, 42:559–557, 1995.

[23] Bush, George W. "National Security Presidential Directive/NSPD-44", 07 December 2005.

[24] Bush, George W. "President's Address to the Nation". http://georgewbush-whitehouse.archives.gov/news/releases/2007/01/20070110-7.html, January 10 2007. Accessed: 2010-09-30.

[25] Cassimatis, Nicholas L. *Harnessing Multiple Representations for Autonomous Full-Spectrum Political, Military, Economic, Social, Information and Infrastructure (PMESII) Reasoning*. Technical report, Air Force Research Lab-Rome, 2007.

[26] Central Intelligence Agency. "Iraq. In The world factbook", 2014. URL https://www.cia.gov/library/publications/the-world-factbook/geos/iz.html.

[27] Chaney, A. D., R. F. Deckro, and J. T. Moore. "Scheduling reconstruction operations with modes of execution", *Journal of the Operational Research Society*, 64(6):898–911, Jun 2013.

[28] Chang, Hyeong S. "On-line sampling-based control for network queueing problems", 2001. URL http://search.proquest.com/docview/304724389?accountid=26185.

[29] Chen, Hsi-Mei. "A non-linear inverse Lanchester square law problem in estimating the force dependent attrition coefficients", *European Journal of Operational Research*, 182:911 – 922, 2007.

[30] Clemens, S. "The Application of Lanchester Models to the Battle of Kursk", 1997. Unpublished Manuscript.

[31] Collier, Craig A. "Two cheers for lethal operations: Money doesn't always achieve what force can", *Armed Forces Journal*, 148(1):41–44, July-August 2010.

[32] Deitchman, S. J. "A Lanchester Model of Guerrilla Warfare", *Operations Research*, 10(6):818–827, 1962.

[33] DeNardo, E.V. *Dynamic Programming: Models and Applications*. Dover Publications, 1982.

[34] Department of Defense. *Doctrine for the Armed Forces of the United States*, JP 1-0. Washington: HQ USA, 20 March 2009.

[35] Department of Defense. *Joint Operations*, JP 3-0. Washington: HQ USA, 22 March 2010.

[36] Department of the Army. *Operations*, FM 3-0. Washington: HQ USA, February 2008.

[37] Dewoody, R., J. Hayes, N. Isnardi, J. Kalinka, and L.T.C.E. Teague. "Irregular warfare models for regional stability development". *Systems and Information Engineering Design Symposium (SIEDS), 2014*, 204–209. April 2014.

[38] Dobbins, James. *The beginner's guide to nation-building*. Rand Corporation, 2007.

[39] Dziedzic, Michael, Barbara Sotrin, and John Agoglia. *Measuring Progress in Conflict Environments (MPICE)- A metrics framework for assessing conflict transitions and stabilization*. Technical report, US Institute for Peace, 2008.

[40] Fox, William P. "Discrete Combat Models: Investigating the Solutions to Discrete Forms of Lanchester's Combat Models", *IJORIS*, 16–34, 2010.

[41] Fricker, R.D. "Attrition Models of the Ardennes Campaign", *Naval Research Logistics*, 45:1–22, 1998.

[42] Goldberg, Jack and Merle C. Potter. *Differential Equations: A Systems Approach.* Upper Saddle River, NJ: Prentice-Hall, 1998.

[43] Goodson, Justin C., Jeffrey W. Ohlmann, and Barrett W. Thomas. "Rollout Policies for Dynamic Solutions to the Multivehicle Routing Problem with Stochastic Demand and Duration Limits", *Operations Research*, 61(1):138–154, 2013.

[44] Hartley, Dean. "DIME/PMESII TOOLS:Past, Present and Future". `http://www.mors.org/UserFiles/file/meetings/08es/hartley_dime.pdf`, 2008.

[45] Helmbold, Robert L. "The Constant Fallacy: A persistent logical flaw in applications of Lanchester's Equations", *European Journal of Operational Research*, 75:647–658, 1994.

[46] III, Marvin L. King. *Optimizing Counterinsurgency Operations.* Ph.D. thesis, Colorado School of Mines, 2014.

[47] Johnson, Dominic D.P. and Joshua S. Madin. "Population Models and Counterinsurgency Strategies". Raphael D. Sagarin and Terrence Taylor (editors), *National Security: A Darwinian Approach to a Dangerous World*, chapter 11, 159–185. Berkeley, CA: University of California Press, 2008.

[48] Jun, D. and D.L. Jones. "The value of sleeping: A rollout algorithm for sensor scheduling in HMMs". *Global Conference on Signal and Information Processing (GlobalSIP), 2013 IEEE*, 181–184. Dec 2013.

[49] Kano, Hiroyuki, Hiroaki Nakata, and Clyde F. Martin. "Optimal curve fitting and smoothing using normalized uniform B-splines: a tool for studying complex systems", *Applied Mathematics and Computation*, 169(1):96 – 128, 2005. ISSN 0096-3003.

[50] Khalil, Hassan K. *Nonlinear Systems.* Upper Saddle River, NJ: Prentice Hall, 2002.

[51] Kleczowski, A. and B. Grenfell. "Mean-Field type equations for spread of epidemics: the 'small world' model", *Physica A*, 275:355–260, 1999.

[52] Kleinman, D., P. Luh, X. Miao, and D. A. Castan. "Distributed Stochastic Resource Allocation in Teams", *IEEE Transactions on Systems, Man and Cybernetics*, 21:61–70, 1991.

[53] Kott, A. and G. Citrenbaum. *Estimating Impact: A Handbook of Computational Methods and Models for Anticipating Economic, Social, Political and Security Effects in International Interventions.* Springer, 2010.

[54] Kress, Moshe and Roberto Szechtman. "Why Defeating Insurgencies Is Hard: The Effect of Intelligence in Counterinsurgency OperationsA Best-Case Scenario", *Operations Research*, 57(3):578–585, 2009.

[55] Lanchester, Frederick W. *Aircraft in Warfare: The Dawn of the Fourth Arm.* London: Constable and Company Limited, 1916.

[56] Lucas, T.W. and T. Turkes. "Fitting Lanchester Equations to the Battle of Kursk and Ardennes", *Naval Research Logistics*, 51:95–116, 2004.

[57] Lukens, M.W. *Strategic Analysis of Irregular Warfare*, USAWC strategy research project. U.S. Army War College, 2010.

[58] Mansoor, Faisal, AbbasK. Zaidi, Lee Wagenhals, and AlexanderH. Levis. "Meta-modeling the Cultural Behavior Using Timed Influence Nets". *Social Computing and Behavioral Modeling*, 1–9. Springer US, 2009.

[59] Marquis, Jefferson P. and Arroyo Center. *Developing an Army strategy for building partner capacity for stability operations / Jefferson P. Marquis ... [et al.].* RAND Arroyo Center Santa Monica, CA, 2010.

[60] Massei, Marina, Alberto Tremori, Francesca Madeo, and Federico Tarone. "Simulation of an Urban Environment by using Intelligent Agents within Asymmetric Scenarios for Assessing Alternative Command & Control Netcentric Maturity Models", *The Journal of Defense Modeling and Simulation: Applications, Methodology, Technology*, 2013.

[61] Mastin, A. and P. Jaillet. May 2014.

[62] McGrew, James S., Jonathan P. How, Brian Williams, and Nicholas Roy. "Air-Combat Strategy Using Approximate Dynamic Programming", *Journal of Guidance, Control, and Dynamics*, 33(5):1641–1654.

[63] Minami, Nathan A. and Paul Kucik. "Developing a Dynamic Model of the Iraqi Insurgency". *Proceedings of the 2009 Spring Simulation Multiconference*, SpringSim '09, 58:1–58:8. Society for Computer Simulation International, 2009.

[64] MINITAB. *version 15.1.1.0.* MINITAB Inc., State College, PA, 2007.

[65] Nabil, S., N. Darwish, and M. Zaki. "A Fairness Based Framework to Resolve Political Disputes", *International Journal of Computer and Information Technology*, 2(3):424–431, May 2013.

[66] Nardo, Michaela, Michaela Saisana, Andrea Saltelli, Stefano Tarantola, Anders Hoffmann, and Enrico Giovannini. *Handbook on Constructing Composite Index Indicators: Methodology and User Guide.* Organisation for Economic Cooperation and Development, 2008.

[67] Novoa, Clara and Robert Storer. "An approximate dynamic programming approach for the vehicle routing problem with stochastic demands", *European Journal of Operational Research*, 196(2):509 – 515, 2009.

[68] Nysether, Nathan E. *Classifying Failing States.* Master's thesis, Air Force Institute of Technology, 2007.

[69] Obama II, Barack H. "National Security Strategy", May 2010.

[70] O'Hanlon, Michael and Ian Livingston. *Iraq Index: Tracking Variables of Reconstruction & Security in Post-Saddam Iraq.* Saban Center for Middle East Policy, November 2011.

[71] O'Hanlon, Michael and Ian Livingston. *Iraq Index: Tracking Variables of Reconstruction & Security in Post-Saddam Iraq.* Saban Center for Middle East Policy, November 2011.

[72] Opper, M. and D. Saad. *Advanced Mean Field Methods: Theory and Practice.* Cambridge, MA: MIT Press, 2001.

[73] Pei, M., S. Kasper, and Carnegie Endowment for International Peace. *Lessons from the Past: The American Record on Nation Building*, Policy brief. Carnegie Endowment for International Peace, 2003.

[74] Persson, P., I. Clasesson, and S. Nordebo. "An adaptive filtering algorithm using mean field annealing techniques", *Proceedings of the IEEE Signal Processing Systems Workshop*, October 2002.

[75] Phillips, Colleen, John Crosscope, and Norman Geddes. "Bayesian Modeling using Belief Networks of Perceived Threat levels Affected by Stratagemical Behavior Patterns". *Proceedings of the Second International Conference on Comutational Cultural Dynamcis*, 55–64. 2008.

[76] Pierson, B. "A System Dynamics model of the FM 3-24 COIN Manual". *Proceedings of the 76th Military Operations Research Symposium*, MORS '08. Military Operations Research Society, 2008.

[77] Richardson, Damon B., Richard F. Deckro, and Victor D. Wiley. "Modeling and Analysis of Post-Conflict Reconstruction", *Journal of Defense Modeling and Simulation*, 1(4):201–213, October 2004.

[78] Richardson, Lewis. *Arms and Insecurity.* Chicago: Quadrangle Books, 1960.

[79] Robbins, Matthew JD. *Investigating the Complexities of Nationbuilding: A Sub-National Regional Perspective*. Master's thesis, Air Force Institute of Technology, 2005.

[80] Robinson II, Paul D. *Patterns of War Termination: A Statistical Approach*. Master's thesis, Air Force Institute of Technology, 2007.

[81] Saie, Cade M. and Darryl K. Ahner. "Investigating the Dynamics of Nation Building Through a System of Differential Equations", 2013. Article submitted for publication.

[82] Sandefur, James T. *Discrete Dynamical Systems: Theory and Applications*. Oxford: Clarendon Press, 1990.

[83] Schaffer, Marvin Baker. "A model of 21st century counterinsurgency warfare", *The Journal of Defense Modeling and Simulation: Applications, Methodology, Technology*, 4(3):252–261, 2007.

[84] Schaffer, M.B. "Lanchester Models of Guerrilla Engagements", *Operations Research*, 457–488, May-June 1968.

[85] Schramm, Harrison C. and Donald P. Gaver. "Lanchester for cyber: The mixed epidemic-combat model", *Naval Research Logistics (NRL)*, 60(7):599–605, 2013.

[86] Secomandi, Nicola. "A Rollout Policy for the Vehicle Routing Problem with Stochastic Demands", *Operations Research*, 49(5):796–802, 2001.

[87] Shearer, Robert and Brett Marvin. "Recognizing Patterns of Nation-State Instability that Lead to Conflict", *Military Operations Research*, 15(3):17–30, 2010.

[88] Smith, Lance. "Military Support to Stabilization, Security, Transition, and Reconstruction Operations Joint Operating Concept", December 2006.

[89] Snieder, Roel. "The role of nonlinearity in inverse problems", *Inverse Problems*, 14:387–404, 1998.

[90] Spliet, Remy, Adriana F. Gabor, and Rommert Dekker. "The vehicle rescheduling problem", *Computers & Operations Research*, 43(0):129 – 136, 2014.

[91] Tannehill, Bryan R. *Forecasting Instability Indicators in the Horn of Africa Region*. Master's thesis, Air Force Institute of Technology, 2008.

[92] Tarantola, Albert and Bernard Valette. "Generalized Nonlinear Inverse Problems Solved Using the Least Squares Criterion", *Reviews of Geophysics and Space Physics*, 20:219–232, 1982.

[93] Tauer, Gregory, Reksh Nagi, and Moises Sudit. "A Dynamic Programming Approach for Nation Building Problems". *Industrial Engineering Research Conference*. 2011.

[94] Taylor, G., R. Bechtel, K. Knudsen, E. Waltz, and J. White. "PSTK: A Toolkit for Modeling Dynamic Power Structures". http://www.soartech.com/images/uploads/file/PSTK-BRIMS-2008-vFINAL.pdf, 2008.

[95] The United States Senate. "The United States Senate: Legislation and Records". http://www.senate.gov/pagelayout/legislative/g_three_sections_with_teasers/legislative_home.htm, February 2012.

[96] The World Bank. "World DataBank". http://databank.worldbank.org/data/home.aspx, 2012.

[97] Turkes, Turker. *Fitting Lanchester and Other Equations to the Battle of Kursk Data*. Master's thesis, Naval Postgraduate School, Monterey, CA, Mar 2000.

[98] Vieira, H., S. M. Sanchez, K. H. Kienitz, and M. C. N. Belderrain. "Efficient, nearly orthogonal-and-balanced, mixed designs", *J Simulation*, 7(4):264–275, Nov 2013.

[99] Vieira, Jr., H. "NOB_Mixed_512DP_template_v1.xls design spreadsheet", March 2012. URL Available online at http://harvest.nps.edu/.

[100] Willard, D. *Lanchester as Force in History: An Analysis of Land Battles of the Years 1618-1905*. Research report, Research Analysis Corp, McLean, VA, Nov 1962.

[101] Wu, Cynara C. *Dynamic resource allocation in CDMA cellular communications systems*. Ph.D. thesis, Massachusetts Institute of Technology, 1999. URL http://hdl.handle.net/1721.1/9332.

[102] Yamamoto, Kazuya. "Nation-building and Integration Policy in the Philippines", *Journal of Peace Research*, 44(2):195–213, March 2007.

[103] Zhang, Canrong, Tao Wu, Kap Hwan Kim, and Lixin Miao. "Conservative allocation models for outbound containers in container terminals", *European Journal of Operational Research*, 238(1):155 – 165, 2014.